Kinematic Control of Redundant Robot Arms Using Neural Networks

Kinematic Control of Redundant Robot Arms Using Neural Networks

Shuai Li
Hong Kong Polytechnic University

Long Jin
Hong Kong Polytechnic University

Mohammed Aquil Mirza
Hong Kong Polytechnic University

Registered Offices
John Wiley & Sons, Inc., 111 River Street, Hoboken, NJ 07030, USA
John Wiley & Sons Ltd, The Atrium, Southern Gate, Chichester, West Sussex, PO19 8SQ, UK

Editorial Office
The Atrium, Southern Gate, Chichester, West Sussex, PO19 8SQ, UK

For details of our global editorial offices, customer services, and more information about Wiley products visit us at www.wiley.com.

Wiley also publishes its books in a variety of electronic formats and by print-on-demand. Some content that appears in standard print versions of this book may not be available in other formats.

Library of Congress Cataloging-in-Publication Data

Names: Li, Shuai, 1983– author. | Jin, Long, 1988– author. |
  Mirza, Mohammed Aquil, 1986– author.
Title: Kinematic control of redundant robot arms using neural networks /
  Shuai Li, Hong Kong Polytechnic University, Long Jin, Hong
  Kong Polytechnic University, Mohammed Aquil Mirza, Hong Kong Polytechnic
  University.
Description: First edition. | Hoboken, NJ : John Wiley & Sons, Inc., 2019. |
  "Most of the materials of this book are derived from the authors' papers
  published in journals and proceedings of the international
  conferences"–Introduction. | Includes bibliographical references and
  index. |
Identifiers: LCCN 2018048730 (print) | LCCN 2018050945 (ebook) | ISBN
  9781119556985 (Adobe PDF) | ISBN 9781119556992 (ePub) | ISBN 9781119556961
  (hardcover)
Subjects: LCSH: Robots–Kinematics–Data processing. | Manipulators
  (Mechanism)–Automatic control. | Redundancy (Engineering)–Data
  processing. | Neural networks (Computer science)
Classification: LCC TJ211.412 (ebook) | LCC TJ211.412 .K563 2019 (print) |
  DDC 629.8/95632–dc23
LC record available at https://lccn.loc.gov/2018048730

Cover Design: Wiley
Cover Image: © alashi/iStock.com

Set in 10/12pt Warnock by SPi Global, Pondicherry, India
Printed in Singapore by C.O.S. Printers Pte Ltd

10 9 8 7 6 5 4 3 2 1

*To our parents and
ancestors, as always*

Contents

List of Figures

List of Tables

Preface

In recent decades, robotics has attracted more and more attention from researchers since it has been widely used in scientific research and engineering applications, such as space exploration, underwater surveys, industrial and military industries, welding, painting and assembly, medical applications, and so on. Much effort has been spent on robotics, and different types of robot manipulators have thus been developed and investigated, such as serial manipulators consisting of redundant manipulators and mobile manipulators, parallel manipulators, and cable-driven manipulators. A redundant manipulator is often designed as a series of links connected by motor-actuated joints that extends from a fixed base to an end-effector while a mobile manipulator is often designed as a robotic device composed of a mobile platform and a redundant manipulator fixed to the platform. Different from these serial manipulators, a parallel manipulator is a mechanical system that usually uses several serial chains to support a single platform, or end-effector. Using these manipulators to save labor and increase accuracy is becoming common practice in various industrial fields. As a consequence, many approaches have been proposed, investigated and employed for the control of robot manipulators. Among them, thanks to the many advantages in parallel distributed structure, nonlinear mapping, ability to learn from examples, high generalization performance, and capability to approximate an arbitrary function with sufficient number of neurons, the neural-network-based approach is a competitive way to control movements of robot manipulators.

In this book, focusing on robot arm control aided with neural networks, we present and investigate different methods and schemes for the control of robot arms. The idea for this book on the redundancy resolution of robot manipulators via different methods and schemes was conceived during research discussion in the laboratory and at international scientific meetings. Most of the material of this book is derived from the authors' papers published in journals and proceedings of international conferences. In fact, in recent decades, the field of robotics has undergone phases of exponential growth, generating many new theoretical concepts and applications (including those of the authors). Our first priority is thus to cover each central topic in enough detail to make the material clear and coherent; in other words, each part (and even each chapter) is written in a relatively self-contained manner.

This book contains 10 chapters which are classified into the following three parts.

Part I: Neural Networks for Serial Robot Arm Control (Chapter 1 through Chapter 6).
Part II: Neural Networks for Parallel Robot Control (Chapter 7 and Chapter 8).
Part III: Neural Networks for Cooperative Control (Chapter 9 and Chapter 10).

Chapter 1 – This chapter breaks these limitations by proposing zeroing neural network (ZNN) models, allowing nonconvex sets for projection operations in activation functions and incorporating new techniques for handing inequality constraint arising in optimizations. Theoretical analyses reveal that the presented ZNN models are of global stability with timely convergence. Finally, illustrative simulation examples are provided and analyzed to substantiate the efficacy and superiority of the presented ZNN models for real-time dynamic quadratic programming subject to equality and inequality constraints.

Chapter 2 – Variable structure strategy is widely used for the control of sensor-actuator systems modeled by Euler–Lagrange equations. However, accurate knowledge on the model structure and model parameters are often required for the control design. In this chapter, we consider model-free variable structure control of a class of sensor–actuator systems, where only the online input and output of the system are available while the mathematic model of the system is unknown. The problem is formulated from an optimal control perspective and the implicit form of the control law is analytically obtained by using the principle of optimality. The control law and the optimal cost function are explicitly solved iteratively. Simulations demonstrate the effectiveness and the efficiency of the proposed method.

Chapter 3 – This chapter identifies two limitations of existing recurrent neural network solutions for manipulator control, i.e., position error accumulation and the convex restriction on the projection set, and overcomes them by proposing two modified neural network models. Similar to the model presented in Chapter 1, the method investigated in this chapter allows nonconvex sets for projection operations, and control error does not accumulate over time in the presence of noise. Unlike most works in which recurrent neural networks are used to process time sequences, the presented approach is model-based and training-free, which makes it possible to achieve fast tracking of reference signals with superior robustness and accuracy. Theoretical analysis reveals the global stability of a system under the control of the presented neural networks. Simulation results confirm the effectiveness of the presented control method in both the position regulation and tracking control of redundant PUMA 560 manipulators.

Chapter 4 – In this chapter, we propose a novel model-free dual neural network, which is able to address the learning and control of manipulators simultaneously in a unified framework. Different from pure learning problems, the interplay of the control part and the learning part allows us to inject an additive noise into the control channel to increase the richness of signals for the purpose of efficient learning. Due to a deliberate design, the learning error is guaranteed for convergence to zero despite the existence of additive noise for stimulation. Theoretical analysis reveals the global stability of the proposed neural network control system. Simulation results verify the effectiveness of the proposed control scheme for redundancy resolution of a PUMA 560 manipulator.

Chapter 5 – In this chapter, we propose a novel recurrent neural network to resolve the redundancy of manipulators for efficient kinematic control in the presence of noises in a polynomial type. Leveraging the high-order derivative properties of polynomial noises, a deliberately devised neural network is presented to eliminate the impact of noises and recover the accurate tracking of desired trajectories in workspace. Rigorous analysis shows that the presented neural law stabilizes the system dynamics and the position tracking error converges to zero in the presence of noises. Extensive simulations verify the theoretical results. Numerical comparisons show that existing dual neural solutions lose stability when exposed to large constant noises or time-varying noises. In contrast, the presented approach works well and has a low tracking error comparable with noise-free situations.

Chapter 6 – In this chapter, we make progress on real-time manipulability optimization by establishing a dynamic neural network for recurrent calculation of manipulability-maximal control actions for redundant manipulators under physical constraints in an inverse-free manner. By expressing position tracking and matrix inversion as equality constraints, physical limits as inequality constraints, and velocity-level manipulability measure, which is affine to the joint velocities, as the objective function, the manipulability optimization scheme is further formulated as a constrained quadratic program. Then, a dynamic neural network with rigorously provable convergence is constructed to solve such a problem online. Computer simulations are conducted and show that, compared with the existing methods, the proposed scheme can raise the manipulability almost 40% on average, which substantiates the efficacy, accuracy and superiority of the proposed manipulability optimization scheme.

Chapter 7 – In this chapter, the kinematic control problem of Stewart platforms is formulated to a constrained quadratic programming. The Karush–Kuhn–Tucker conditions of the problem are obtained by considering the problem in its dual space, and then a dynamic neural network is designed to solve the optimization problem recurrently. Theoretical analysis reveals the global convergence of the employed dynamic neural network to the optimal solution in terms of the defined criteria. Simulation results verify the effectiveness in the tracking control of the Stewart platform for dynamic motions.

Chapter 8 – In this chapter, we establish a model-free dual neural network to control the end-effector of a Stewart platform for the tracking of a desired spatial trajectory, at the same time as learning the unknown time-varying parameters. The proposed model is purely data driven. It does not rely on the system parameters as *a priori* and provides a new solution for stabilization of the self motion of Stewart platforms. Theoretical analysis and results show that we can achieve a globally convergent neural model in this chapter. It is also shown to be optimal per the model-free criterion. In this chapter, numerical simulations are those which highlight and illustrate relatable performance capability in terms of model-free optimization. Simulation results provided verify the tracking control of the end-effector efficiently while controlling the dynamic motion of the Stewart platform.

Chapter 9 – In this chapter, a distributed scheme is proposed for the cooperative motion generation in a distributed network of multiple redundant manipulators. The proposed scheme can simultaneously achieve the specified primary task to reach global cooperation under limited communications among manipulators and optimality in terms of a specified optimization index of redundant robot manipulators. The proposed distributed scheme is reformulated as a quadratic program (QP). To inherently suppress noises originating from communication interferences or computational errors, a noise-tolerant zeroing neural network (NTZNN) is constructed to solve the QP problem online. Then, theoretical analyses show that, without noise, the proposed distributed scheme is able to execute a given task with exponentially convergent position errors. Moreover, in the presence of noise, the proposed distributed scheme with the aid of the NTZNN model has a satisfactory performance. Furthermore, simulations and comparisons based on PUMA 560 redundant robot manipulators substantiate the effectiveness and accuracy of the proposed distributed scheme with the aid of the NTZNN model.

Chapter 10 – This chapter investigates the distributed motion planning of multiple robot arms with limited communications in the presence of noises. To do this, a nonlinearly activated noise-tolerant zeroing neural network (NANTZNN) is designed and presented for the first time for solving the presented distributed scheme online. Theoretical analyses and simulation results show the effectiveness and accuracy of the presented distributed scheme with the aid of the NANTZNN model.

This book is written for academic and industrial researchers as well as graduate students studying in the developing fields of robotics, numerical algorithms, and neural networks. It provides a comprehensive view of the combined research of these fields, in addition to the accomplishments, potentials, and perspectives. We hope that this book will generate curiosity and interest for those wishing to learn more, and that it will provide new challenges to seek new theoretical tools and practical applications. Without doubt, this book can be extended. Any comments or suggestions are welcome. The authors can be contacted via e-mail: shuaili@polyu.edu.hk; jinlongsysu@foxmail.com; and csmaquil@comp.polyu.edu.hk.

Hong Kong, 2018   *Shuai Li, Long Jin and Mohammed Aquil Mirza*

Acknowledgments

This book basically comprises the results of many original research papers of the authors' research group, in which many authors of these original papers have done a great deal of detailed and creative research work. Therefore, we are much obliged to our contributing authors for their high-quality work.

We acknowledge the continuous support of our research by the National Natural Science Foundation of China (No. 61401385), the Fundamental Research Funds for the Central Universities (No. lzujbky-2017-37), the Hunan Natural Science Foundation of China (No. 2017JJ3257), the Hong Kong Research Grants Council Early Career Scheme (No. 25214015), the Hong Kong Polytechnic University (Nos G-YBMU, G-UA7L, 4-ZZHD, F-PP2C, and 4-BCCS), and also by the Research Foundation of Education Bureau of Hunan Province, China (No. 17C1299). In addition we would like to thank the editors sincerely for their time and efforts spent in handling this book, as well as for their constructive comments and suggestions. We are very grateful to the staff at Wiley and IEEE for their invaluable support during the preparation and publication of this book.

To all these wonderful people we owe a deep sense of gratitude especially now that the research projects and the book have been completed.

Part I

Neural Networks for Serial Robot Arm Control

1

Zeroing Neural Networks for Control

1.1 Introduction

In addition to the remarkable features such as parallelism, distributed storage, and adaptive self-learning capability, neural networks can be readily implemented by hardware, and have thus been applied widely in many fields [1–6]. The zeroing neural network (ZNN) as well as its variant (i.e. zeroing dynamic), as a systematic approach to the online solution of time-varying problems with scalar situation included, has been applied to online matrix inversion [7], motion generation and control of redundant robot manipulators [3], and tracking control of nonlinear chaotic systems [8]. For example, a ZNN model with a nonlinear function activated is applied to the kinematic control of redundant robot manipulators via Jacobian matrix pseudoinversion in [9], which achieves high accuracy but cannot handle the bound constraints existing in the robots. In [10], present a finite-time convergent ZNN model is presented for solving dynamic quadratic programs with application to robot tracking, which requires convex activation functions and cannot remedy the issue of joint-limit avoidance. Such a ZNN method is further discretized to compute the solution to time-varying nonlinear equations based on a new three-step formula, which can be implemented on a digital computer directly. In addition, for the applications, ZNN is exploited in [3] to remedy the joint-angle drift phenomenon of redundant robot manipulators by minimizing the difference between the desired joint position and the actual one.

It is worth pointing out here that although these existing models differ in choosing different error functions or using different activation functions, all of them follow similar design procedures: the ZNN method usually formulates a time-varying problem into a regulation problem in control. Specifically, the residual error of a ZNN model for the task function to be solved is to be regulated to zero. Then, a monotonically increasing and odd function activated ZNN model with its equilibrium identical to the solution of this time-varying problem is devised to solve the latter recursively. In addition, the design parameter in the ZNN method should be larger than zero. To the best of the authors' knowledge, all existing results on ZNN assume that the set for projection of the activation function is a convex one, which evidently excludes the nonconvex set from consideration. General conclusions relaxing the convex constraint on activation functions remain unexplored.

Kinematic Control of Redundant Robot Arms Using Neural Networks, First Edition.
Shuai Li, Long Jin and Mohammed Aquil Mirza.
© 2019 John Wiley & Sons Ltd. Published 2019 by John Wiley & Sons Ltd.

In this chapter, we make progress in this direction by proposing new results on ZNN to remedy these weaknesses. The presented ZNN models in this chapter are able to deal with nonconvex projection set Ω in the activation functions, while the existing solutions require the projection set to be convex. Additionally, this is the first work on ZNN for solving a time-varying optimization problem with inequality and bound constraints, which opens a door to the research on solving time-varying constrained optimization problems in an error-free manner. In short, there are two limitations in the existing research on ZNN, i.e. lacking the technique for handling inequality and bound constraints when solving dynamic optimization problems and requiring the activation function to be odd and monotonically increasing. This chapter overcomes these limitations by proposing ZNN models, allowing nonconvex sets for projection operations in activation functions and incorporating new techniques for handling inequality constraint.

1.2 Scheme Formulation and ZNN Solutions

In this section, a ZNN model for dynamic quadratic programming subject to equality and inequality constraints is presented. Then, new results are derived using the ZNN model with the aid of a nonconvex activation function.

1.2.1 ZNN Model

Consider the convex dynamic quadratic programming subject to equality and inequality constraints in the form of

$$\text{minimize} \quad x^T(t)P(t)x(t)/2 + q^T(t)x(t),$$
$$\text{subject to} \quad A(t)x(t) = b(t),$$
$$C(t)x(t) \leq d(t), \tag{1.1}$$

where superscript ^{T} denotes the transpose operation over a vector or a matrix; smoothly time-varying matrix $P(t) \in \mathbb{R}^{n\times n}$ is positive-definite; $q(t) \in \mathbb{R}^n$, $A(t) \in \mathbb{R}^{m\times n}$ being of full-row-rank, $C(t) \in \mathbb{R}^{p\times n}$ and $b(t) \in \mathbb{R}^m$, $d(t) \in \mathbb{R}^p$ are all smoothly time-varying.

By adding a time-varying nonnegative term to the inequality constraint, the convex dynamic quadratic programming problem (1.1) is converted into

$$\text{minimize} \quad x^T(t)P(t)x(t)/2 + q^T(t)x(t),$$
$$\text{subject to} \quad A(t)x(t) - b(t) = 0,$$
$$C(t)x(t) - d(t) + \bar{\sigma}^2(t) = 0, \tag{1.2}$$

where $\bar{\sigma}^2(t) \in \mathbb{R}^p$ is defined as $\bar{\sigma}^2(t) = \sigma(t)\circ\sigma(t) = [\sigma_i^2(t)]$. Define a Lagrange function as follows:

$$L(x(t), \rho_1(t), \rho_2(t), \sigma(t), t) = x^T(t)P(t)x(t)/2$$
$$+ q^T(t)x(t) + \rho_1^T(t)(A(t)x(t) - b(t))$$
$$+ \rho_2^T(t)(C(t)x(t) - d(t) + \bar{\sigma}^2(t)). \tag{1.3}$$

By using the related Karush–Kuhn–Tucker condition [11], we have

$$P(t)x(t) + q(t) + A^{\mathrm{T}}(t)\rho_1(t) + C^{\mathrm{T}}(t)\rho_2(t) = 0,$$
$$A(t)x(t) - b(t) = 0,$$
$$C(t)x(t) - d(t) + \bar{\sigma}^2(t) = 0,$$
$$\rho_2(t) \circ \sigma(t) = 0. \tag{1.4}$$

Letting $y(t) = [x(t), \rho_1(t), \rho_2(t), \sigma(t)]^{\mathrm{T}}$, the above equation can be rewritten as

$$f(y(t), t) = 0 \in \mathbb{R}^{m+n+2p}, \tag{1.5}$$

where mapping function $f(\cdot)$ is used to denote the left-hand side of (1.4). By defining $\varepsilon(t) = -f(y(t), t)$, we adopt the following evolution for $\varepsilon(t)$, i.e. the ZNN design formula:

$$\dot{\varepsilon}(t) = -\gamma\varepsilon(t). \tag{1.6}$$

Substituting (1.5) into (1.6) yields a dynamic equation:

$$J(y(t), t)\dot{y}(t) = -\gamma(f(y(t), t)) - \frac{\partial f(y(t), t)}{\partial t}, \tag{1.7}$$

where

$$J(y(t), t) = \frac{\partial f(y(t), t)}{\partial y(t)} = \begin{bmatrix} P(t) & A^{\mathrm{T}}(t) & C^{\mathrm{T}}(t) & 0 \\ A(t) & 0 & 0 & 0 \\ C(t) & 0 & 0 & \bar{\sigma}(t) \\ 0 & 0 & 2\bar{\sigma}(t) & \bar{\rho}_2(t) \end{bmatrix} \in \mathbb{R}^{(m+n+2p)\times(m+n+2p)},$$

with

$$\bar{\sigma}(t) = \begin{bmatrix} \sigma_1(t) & 0 & \cdots & 0 \\ 0 & \sigma_2(t) & \cdots & 0 \\ \vdots & \vdots & \ddots & \vdots \\ 0 & 0 & 0 & \sigma_p(t) \end{bmatrix}, \quad \bar{\rho}_2(t) = \begin{bmatrix} \rho_1(t) & 0 & \cdots & 0 \\ 0 & \rho_2(t) & \cdots & 0 \\ \vdots & \vdots & \ddots & \vdots \\ 0 & 0 & 0 & \rho_p(t) \end{bmatrix}.$$

For the situation of Jacobian matrix $J(x(t), t)$ being nonsingular, the above equation is further rewritten as

$$\dot{y}(t) = -\gamma J^{-1}(y(t), t)\left(f(y(t), t) - \frac{\partial f(y(t), t)}{\partial t}\right), \tag{1.8}$$

where $y(t)$, starting from a given initial condition, denotes the neural state as well as the output corresponding to theoretical solution $y^*(t)$, with its first n elements constituting the optimal solution $x^*(t)$ to (1.1).

Remark 1.1 As stated in [12], the only systematic approach presented so far for the solution of time-varying problems is the ZNN-based technique. Under suitable assumptions, it is shown that the generated solution for solving a time-varying problem synthesized by the ZNN-based technique converges to the exact theoretical time-varying solution globally and exponentially [12]. A gradient-based (or gradient-related) technique is widely used for the online solving of various problems, which is taken as an example to compare with the ZNN-based technique in this remark. The differences between these two techniques are listed in Table 1.1.

Table 1.1 Comparison of ZNN-based and gradient-based techniques for solving $f(y(t), t) = 0$.

	Error function	Design formula	Dynamic equation
ZNN-based technique	$\varepsilon(t)$ $= -f(y(t), t)$	$\dot{\varepsilon}(t) = -\gamma\varepsilon(t)$	$\dot{y}(t) = -\gamma J^{-1}(y(t), t)$ $(f(y(t), t) - \frac{\partial f(y(t),t)}{\partial t})$
Gradient-based technique	$e(t)$ $= \parallel f(y(t), t) \parallel_2^2 / 2$	$\dot{y}(t) = -\gamma \frac{\partial e(t)}{\partial y}$	$\dot{y}(t) = -\gamma J^{T}(y(t), t)$ $f(y(t), t)$

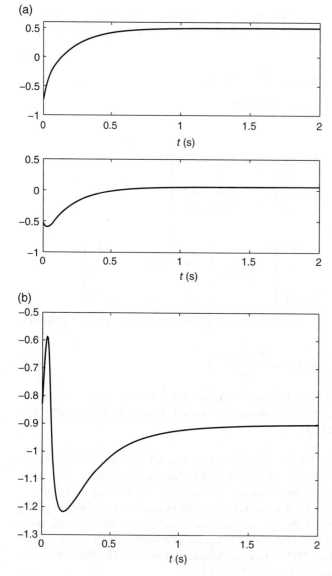

Figure 1.1 State vector $x(t)$, $\rho_1(t)$ of ZNN model (1.8) for solving (1.13) at $t = 1$. (a) Profiles of $x(t)$ and (b) profile of $\rho_1(t)$.

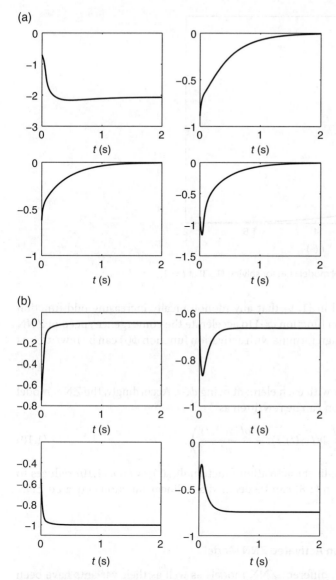

Figure 1.2 State vector $\rho_2(t), \sigma(t)$ of ZNN model (1.8) for solving (1.13) at $t = 1$. (a) Profiles of $\rho_2(t)$ and (b) profile of $\sigma(t)$.

In addition, it is revealed in [13] that a controller designed by the ZNN-based technique is stable naturally as long as design parameter $\gamma > 0$, while a controller designed by other techniques cannot have a guaranteed stability. This can be deemed as another advantage of the ZNN-based technique compared with other existing techniques.

A disadvantage of the ZNN-based technique compared with other existing techniques is that, as shown in Table 1.1, the matrix inversion operation required in the model may lead to the failure of the solving task when encountering a singularity.

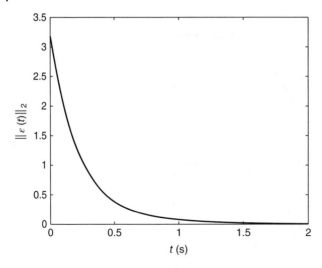

Figure 1.3 Residual error of ZNN model (1.8) for solving (1.13) at $t = 1$.

Remark 1.2 It is revealed in [1, 9] that any monotonically increasing odd function can be used as the activation function $\phi(\cdot)$ to accelerate the convergence speed of ZNN models. Then, the ZNN design formula with activation function $\phi(\cdot)$ can be rewritten as

$$\dot{\varepsilon}(t) = -\gamma\Phi(\varepsilon(t)), \tag{1.9}$$

where $\Phi(\cdot)$ denotes a vector with each element being $\phi(\cdot)$. Accordingly, the ZNN model (1.8) with activation function can be rewritten as

$$\dot{y}(t) = -\gamma J^{-1}(y(t), t)\left(\Phi(f(y(t), t)) - \frac{\partial f(y(t), t)}{\partial t}\right). \tag{1.10}$$

Note that, by exploiting the linear activation function $\Phi(\varepsilon(t)) = \varepsilon(t)$, (1.10) reduces to (1.8). That is to say, Equation (1.8) can be deemed as a linear function activated ZNN model.

1.2.2 Nonconvex Function Activated ZNN Model

As reviewed in Section 1.1, different ZNN models as well as their variants have been extensively studied and exploited for solving dynamic problems over the past 15 years. Although these existing models differ in choosing different error functions or using different activation functions, all of them follow similar design procedures and share the same convergence condition. For example, the ZNN model is often designed as a dynamic system with its equilibrium identical to the solution of the problem to be solved and then solves the latter recursively. In addition, the design parameter γ should be larger than zero and the activation function used to accelerate the convergence speed should be monotonically increasing and odd. To the best of the authors' knowledge, all existing results on ZNN assume that the set for projection of activation function is a convex one, which evidently excludes a nonconvex set from consideration. General conclusions relaxing the convex constraint on activation functions remain unexplored.

Let $\Upsilon_\Omega(U)$ be the projection from a set U to a set Ω such that $\Upsilon_\Omega(U) = \arg\min_{Y \in \Omega}$ $\| Y - U \|_2$ with $0 \in \Omega$, we show the new design formula as follows:

$$\dot{\varepsilon}(t) = -\Upsilon_\Omega(\varepsilon(t)). \tag{1.11}$$

Expanding (1.11) leads to a new nonconvex function activated ZNN model:

$$\dot{y}(t) = -J^{-1}(y(t), t)\left(\Upsilon_\Omega(\gamma f(y(t), t)) - \frac{\partial f(y(t), t)}{\partial t}\right). \tag{1.12}$$

It can be concluded from the definition of $\Upsilon_\Omega(\cdot)$ that $\Upsilon_\Omega(\cdot)$ incorporates the existing ZNN activation functions as special cases. That is to say, any monotonically increasing odd activation function $\phi(\cdot)$ can be deemed as a subcase of $\Upsilon_\Omega(\cdot)$. In addition, it also can be generalized that, different from the existing results, the following special set can be used as the activation function of ZNN.

$\Omega = \{U \in \mathbb{R}^{(n+m+2p)}, -c_1 \le U_i \le c_1 \quad \text{or} \quad U_i = \pm c_2\}$, where c_1 and c_2 are two constants and $c_1 < c_2$.

1.3 Theoretical Analyses

In this section, we conduct analyses on the convergence of the presented ZNN model (1.8) and (1.12) via the following theorems.

Theorem 1.1 The presented ZNN model (1.8) is stable and is exponentially and globally convergent to an equilibrium point $y^*(t)$, of which the first n elements constitute the optimal solution $x^*(t)$ to (1.1).

Proof: In terms of (1.1), ZNN model (1.8) is an equivalent expansion of $\dot{\varepsilon}(t) = -\gamma\varepsilon(t)$. By selecting a Lyapunov function candidate $v(t) = \varepsilon^2(t)$ and by using the Lyapunov theory, one can readily derive that ZNN model (1.8) is stable with $\dot{v}(t) = -2\gamma\varepsilon(t)$.

In addition, solving the dynamic system $\dot{\varepsilon}(t) = -\gamma\varepsilon(t)$ directly, we have $\varepsilon(t) = \varepsilon(0)\exp(-\gamma t)$ with $\varepsilon(0)$ denoting the initial value of $\varepsilon(t)$. Therefore, we come to the conclusion that, starting from any initial condition $\varepsilon(0)$, the residual error $\varepsilon(t)$ of ZNN model (1.8) for solving (1.1) globally and exponentially converges to zero. That is, the state vector $y(t)$ globally and exponentially converges to an equilibrium point $y^*(t)$, of which the first n elements constitute the optimal solution to (1.1). The proof is thus completed. ∎

Theorem 1.2 ZNN model (1.12) globally converges to the theoretical solution of convex dynamic quadratic programming problem (1.1) subject to equality and inequality constraints.

Proof: A Lyapunov function candidate can be defined for (1.11) as

$$V(t) = \varepsilon^T(t)\varepsilon(t)/2,$$

which is positive definite in view of the facts that $V(t) > 0$ for any $\varepsilon(t) \neq 0$, and that $V(t) = 0$ only for $\varepsilon(t) = 0$. In addition, the time derivative of $V(t)$ can be derived as

$$\dot{V}(t) = -\varepsilon^T(t)\Upsilon_\Omega(\varepsilon(t)).$$

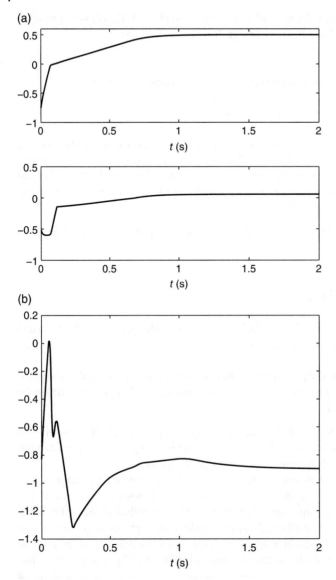

Figure 1.4 State vector $x(t)$, $\rho_1(t)$ of ZNN model (1.8) for solving (1.13) at $t = 1$. (a) Profiles of $x(t)$ and (b) profile of $\rho_1(t)$.

According to the definition of $\Upsilon_\Omega(\varepsilon(t))$, we have

$$\| \Upsilon_\Omega(\varepsilon(t)) - \varepsilon(t) \|_2^2 \leq \| Y - \varepsilon(t) \|_2^2, \quad \forall Y \in \Omega.$$

Choosing $Y = 0$ leads to

$$\| \Upsilon_\Omega(\varepsilon(t)) - \varepsilon(t) \|_2^2 \leq \| \varepsilon(t) \|_2^2,$$

which is equivalent to

$$\Upsilon_\Omega^T(\varepsilon(t))\Upsilon_\Omega(\varepsilon(t)) - 2\Upsilon_\Omega^T(\varepsilon(t))\varepsilon(t) \leq 0.$$

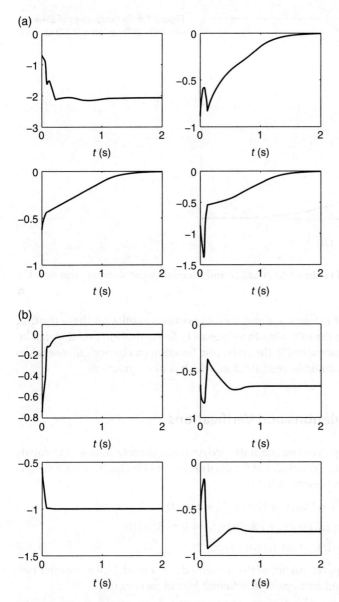

Figure 1.5 State vector $p_2(t)$, $\sigma(t)$ of ZNN model (1.8) for solving (1.13) at $t = 1$. (a) Profiles of $p_2(t)$ and (b) profiles of $\sigma(t)$.

Reformulating the above inequality produces

$$2\Upsilon_\Omega^T(\varepsilon(t))\varepsilon(t) \geq \Upsilon_\Omega^T(\varepsilon(t))\Upsilon_\Omega(\varepsilon(t)) \geq 0.$$

Therefore, we have

$$\dot{V}(t) \leq -\Upsilon_\Omega^T(\varepsilon(t))\Upsilon_\Omega(\varepsilon(t))/2 \leq 0.$$

It can be readily summarized that $\varepsilon(t)$ globally converges to zero. That is, ZNN model (1.12) globally converges to the theoretical solution of convex dynamic quadratic

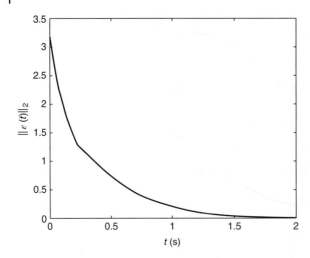

Figure 1.6 Residual error of ZNN model (1.8) for solving (1.13) at $t = 1$.

programming problem (1.1) subject to equality and inequality constraints. The proof is thus complete. ∎

Remark 1.3 Theorem 1.2 extends and generalizes previous results on the activation function of ZNN models in the following three ways. (1) The activation function can be nonsymmetric and nonmonotonic. (2) The activation function can be nondifferentiable. (3) The activation function could have saturation, which is more practical.

1.4 Computer Simulations and Verifications

In this section, the following dynamic quadratic programming problem subject to equality and inequality constraints is considered for illustration and for comparison, which is modified from the problem presented in [1]:

$$\text{minimize} \quad ((\sin t)/4 + 1)x_1^2(t) + ((\cos t)/4 + 1)x_2^2(t)$$
$$+ (\cos t)x_1(t)x_2(t) + (\sin 3t)x_1(t) + (\cos 3t)x_2(t),$$
$$\text{subject to} \quad (\sin 4t)x_1(t) + (\cos 4t)x_2(t) = \cos 2t. \tag{1.13}$$

In order to investigate the performance of the presented ZNN models, we consider two examples in the following subsections with different bound constraints.

1.4.1 ZNN for Solving (1.13) at $t = 1$

At $t = 1$ and with the bound constraint incorporated, (1.13) can be further rewritten as

$$\text{minimize} \quad ((\sin 1)/4 + 1)x_1^2(t) + ((\cos 1)/4 + 1)x_2^2(t)$$
$$+ (\cos 1)x_1(t)x_2(t) + (\sin 3)x_1(t) + (\cos 3)x_2(t),$$
$$\text{subject to} \quad (\sin 4)x_1(t) + (\cos 4)x_2(t) = \cos 2,$$
$$-0.5 \leq x_1 \leq 0.5,$$
$$-0.5 \leq x_2 \leq 0.5. \tag{1.14}$$

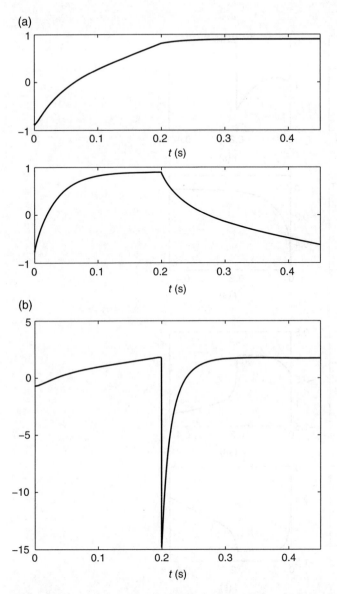

Figure 1.7 State vector $x(t)$, $\rho_1(t)$ of ZNN model (1.12) for solving (1.16). (a) Profiles of $x(t)$ and (b) profile of $\rho_1(t)$.

Starting with a randomly generated initial state, the corresponding computer simulation results are shown in Figures 1.1–1.3. Specifically, the element trajectories of the state $x(t)$, $\rho_1(t)$, $\rho_2(t)$ and $\sigma(t)$ are shown in Figures 1.1 and 1.2, from which we could observe that the solution of ZNN model (1.8) satisfies the given bound constraint. In addition, the corresponding residual error shown in Figure 1.3 further illustrates the effectiveness of the presented ZNN model (1.8).

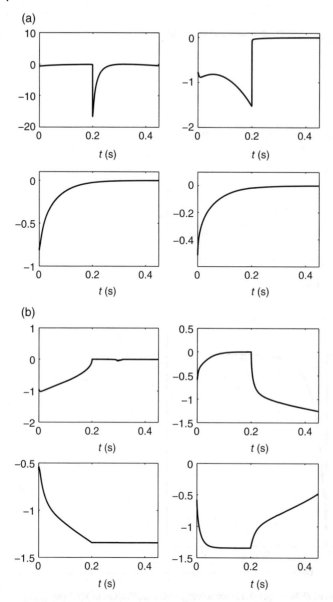

Figure 1.8 State vector $p_2(t)$, $\sigma(t)$ of ZNN model (1.12) for solving (1.16). (a) Profiles of $p_2(t)$ and (b) profiles of $\sigma(t)$.

It is revealed in Theorem 1.2 that the projection operation for activation functions could be nonconvex. In this section, to exemplify the choice of Ω, we particularly consider the following set:

$$\Omega = \{\chi = [\chi_i] \in \mathbb{R}^{m+n+2p}, -c_2 \le \chi_i \le c_2, \text{ or } \chi_i = \pm c_1\}, \tag{1.15}$$

with $c_1 = 1$ and $c_2 = 0.1$ in the simulation. The choice of Ω is nonconvex due to the fact that $0 \in \Omega$ and $1 \in \Omega$ but $(0+1)/2 \notin \Omega$. Physically, Ω defined in (1.15) is generalized

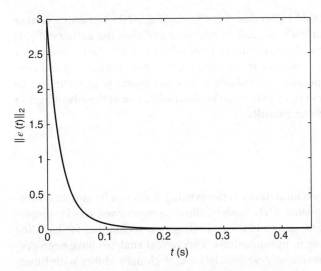

Figure 1.9 Residual error of ZNN model (1.12) for solving (1.16).

from commonly used strategies in industrial bang-bang control, where only maximum input action c_1, minus maximum input action $-c_1$, and zero input action 0 are applicable. To avoid chattering phenomena in conventional bang-bang control, it is preferable to expand zero input action into a small range $[-c_2, c_2]$, which results in the definition of Ω in (1.15). As shown in Figures 1.4–1.6, the state vector $x(t)$ is kept within its bound and the residual error converges to zero as time evolves. The convergence of the residual error in Figure 1.6 validates the effectiveness of Theorem 1.2 for the nonconvex constraint on activation function.

1.4.2 ZNN for Solving (1.13) with Different Bounds

With a new bound constraint incorporated, (1.13) can be further rewritten as

$$\text{minimize} \quad ((\sin t)/4 + 1)x_1^2(t) + ((\cos t)/4 + 1)x_2^2(t)$$
$$+ (\cos t)x_1(t)x_2(t) + (\sin 3t)x_1(t) + (\cos 3t)x_2(t),$$
$$\text{subject to} \quad (\sin 4t)x_1(t) + (\cos 4t)x_2(t) = \cos 2t,$$
$$-0.9 \le x_1 \le 0.9,$$
$$-0.9 \le x_2 \le 0.9. \tag{1.16}$$

Starting with a randomly generated initial state, the corresponding computer simulation results are shown in Figures 1.7–1.9. Specifically, the element trajectories of the state $x(t)$ and $\rho_1(t)$, and $\rho_2(t)$ and $\sigma(t)$ are shown in Figures 1.7 and 1.8, respectively, from which we could observe that the solution of ZNN model (1.8) satisfies the given bound constraint. In addition, all the states vary as time evolves. Besides, the corresponding residual error shown in Figure 1.9 further illustrates the effectiveness of the presented ZNN model (1.8).

Remark 1.4 A ZNN model is investigated in [14] for solving a quadratic programming problem with application to the repetitive motion generation of redundant robot manipulators. However, as stated in Section 1.1, the existing ZNN technique cannot solve an optimization problem with inequality constraint and thus the authors of [14] do not consider the joint physical constraints in their scheme. With the technique for handling inequality constraint presented in this chapter, the motion planning and control of redundant robot manipulators formulated as a constrained QP problem can be solved by ZNN readily and accurately, which can be deemed as one of the advantages of this chapter compared with existing results.

1.5 Summary

This chapter has pointed out two limitations in the existing ZNN results and then overcome these limitations by proposing ZNN models, allowing nonconvex sets for projection operations in activation functions and incorporating new techniques for handling the inequality constraint arising in optimizations. Theoretical analyses have been presented and shown that the presented ZNN models are of global stability with timely convergence. Finally, illustrative simulation examples have been provided and analyzed to substantiate the efficacy and superiority of the presented ZNN models for real-time dynamic quadratic programming subject to equality and inequality constraints. This chapter can be deemed as a rudiment of further investigations on constrained dynamic optimization with time-varying parameters, which can be generalized and employed for the motion planning and control of redundant robot manipulators [15] and distributed winner-take-all-based task allocation of multiple robots [16].

2

Adaptive Dynamic Programming Neural Networks for Control

2.1 Introduction

With the development of mechatronics, automatic systems consisting of sensors for perception and actuators for action are being more and more widely used [17]. Besides the proper choices of sensors and actuators and an elaborate fabrication of mechanical structures, the control law design also plays a crucial role in the implementation of automatic systems especially for those with complicated dynamics. For most mechanical sensor–actuator systems, it is possible to model them in Euler–Lagrange equations [17]. In this chapter, we are concerned with the sensor–actuator systems modeled by Euler–Lagrange equations.

Due to the importance of Euler–Lagrange equations in modeling many real sensor–actuator systems, much attention has been paid to the control of such systems. According to the type of constraints, the Euler–Lagrange system can be categorized as a Euler–Lagrange system without nonholonomic constraints (e.g. fully actuated manipulator [18], omni-directional mobile robot [19]), the under-actuated multiple body system. For a Euler–Lagrange system without nonholonomic constraints, the input dimension are often equal to the output dimensions and the system is often able to be transformed into a double integrator system by employing feedback linearization [20]. Other methods, such as the control Lyapunov function method, passivity-based method, and optimal control method are also successfully applied to control the Euler–Lagrange system without nonholonomic constraints. In contrast, as the input dimensions are lower than those of outputs, it is often impossible to directly transform the Euler–Lagrange system subject to nonholonomic constraints to a linear system and thus feedback linearization fails to stabilize the system. To tackle the problem, various methods (variable structure control-based method [21], backstepping-based control [22], optimal control-based method, and discontinuous control method) have been widely investigated and some useful design procedures are proposed. However, due to the inherent nonlinearity and nonholonomic constraints, most existing methods [21], [22] are strongly model dependent and the performance is very sensitive to model errors. Inspired by the success of human operators for the control of Euler–Lagrange systems, various intelligent control strategies are proposed to solve the control problem of Euler–Lagrange systems subject to nonholonomic constraints. As demonstrated by extensive simulations, these type of strategies are indeed effective for the control of Euler–Lagrange systems subject to nonholonomic constraints. However, rigorous proof of the stability is difficult for this type of method and there may exist some initializations of the state, from which the system cannot be stabilized.

In this chapter, we propose a self-learning control method applicable to Euler–Lagrange systems. In contrast to existing work on intelligent control of Euler–Lagrange systems, the stability of the closed loop system with the proposed method is proven in theory. On the other hand, different from model-based design strategies, such as backstepping-based design [22] and variable structure-based design [21], the proposed method does not require information on the model parameters and therefore is a model independent method. We formulate the problem from an optimal control perspective. In this framework, the goal is to find the input sequence to minimize the cost function defined on infinite horizon under the constraint of the system dynamics. The solution can be found by solving a Bellman equation according to the principle of optimality [23]. Then an adaptive dynamic programming strategy is utilized to numerically solve the input sequence in real time.

2.2 Preliminaries on Variable Structure Control of the Sensor–Actuator System

In this chapter, we are concerned with the following sensor–actuator system in the Euler–Lagrange form,

$$D(q)\ddot{q} + C(q, \dot{q})\dot{q} + \phi(q) = u, \tag{2.1}$$

where $q \in \mathbb{R}^n$, $D(q) \in \mathbb{R}^{n \times n}$ is the inertial matrix, $C(q, \dot{q}) \in \mathbb{R}^{n \times n}$, $\phi(q) \in \mathbb{R}^n$ and $u \in \mathbb{R}^n$. Note that the inertial matrix $D(q)$ is symmetric and positive definite. There are three terms on the left-hand side of Equation (2.1). The first term involves the inertial force in the generalized coordinates, the second one models the Coriolis force and friction, the values of which depend on $\dot{q}$, and the third one is the conservative force, which corresponds to the potential energy. The control force u applied on the system drives the variation of the coordinate q. It is also noteworthy that we assume the dimension of u is equal to that of q here. This definition also admits the case for u with lower dimension than that of q by imposing constraints to u, e.g. the constraint $u = [u_1, u_2, ..., u_n]$ with $u_1 = 0$ restricts u in a $n - 1$ dimensional space. Defining state variables $x_1 = q$ and $x_2 = \dot{q}$, the Euler–Lagrange equation (2.1) can be put into the following state-space form:

$$\dot{x}_1 = x_2$$
$$\dot{x}_2 = -D^{-1}(x_1)(u + C(x_1, x_2)x_2 + \phi(x_1)). \tag{2.2}$$

Note that the matrix $D(x_1)$ is invertible as it is positive definite. The control objective is to asymptotically stabilize the Euler–Lagrange system (2.2), i.e. design a mapping $(x_1, x_2) \to u$ such that $x_1 \to 0$ and $x_2 \to 0$ when time elapses.

As an effective design strategy, variable structure control finds applications in many different types of control systems including the Euler–Lagrange system. The method stabilizes the dynamics of a nonlinear system by steering the state to a elaborately designed sliding surface, on which the state inherently evolves towards the zero state. Particularly for the system (2.2), we define $s = s(x_1, x_2)$ as follows:

$$s = c_0 x_1 + x_2, \tag{2.3}$$

where $c_0 > 0$ is a constant. Note that $s = c_0 x_1 + x_2 = 0$ together with the dynamics of x_1 in Equation (2.2) gives the dynamics of x_1 as $\dot{x}_1 = -c_0 x_1$ for $c_0 > 0$. Clearly, x_1 asymptotically converges to zero. Also we know $x_2 = 0$ when $x_1 = 0$ according to $s = c_0 x_1 + x_2 = 0$. Therefore, we conclude the states x_1, x_2 on the sliding surface $s = 0$ for s defined in Equation (2.3) converge to zero with time. With this property of the sliding surface, a control law driving the states to $s = 0$ definitely guarantees the ultimate convergence to the zero states. Accordingly, the stabilization of the system can be realized by controlling s to zero. To reach this goal, a positive definite control Lyapunov function $V(s)$, e.g. $V(s) = s^2$, is often used to design the control law. For stability consideration, the time derivative of $V(s)$ is required to be negative definite. In order to guarantee the negative definiteness of the time derivative of $V(s)$, exact information about the system dynamics (2.2) is often necessary, which results in the model-based design strategies.

We have the following remark about the Euler–Lagrange equation (2.1) for modeling sensor–actuator systems.

Remark 2.1 In this chapter, we are concerned with the class of sensor–actuator systems modeled by the Euler–Lagrange equation (2.1). Actually, the dynamics of mechanical systems can be described by the Euler–Lagrange equation according to the rigid body mechanics [17], which is essentially equivalent to Newton's laws of motion. Therefore, a mechanical sensor–actuator system can be modeled by Equation (2.1). In this regard, the Euler–Lagrange equation employed in the is chapter models a general class of sensor–actuator systems.

2.3 Problem Formulation

Without losing generality, we stabilize the system (2.1) by steering it to the sliding surface $s = 0$ with s defined in Equation (2.3). Different from existing model-based design procedures, we design a self-learning controller, which does not require accurate knowledge about $D(q)$, $C(q, \dot{q})$, and $\phi(q)$ in Equation (2.1). In this section, we formulate such a control problem from the optimal control perspective.

In this chapter, we set the origin as the desired operating point, i.e. we consider the problem of controlling the state of the system (2.1) to the origin. For the case with other desired operating points, the problem can be equivalently transformed to the one with the origin as the operating point by shifting the coordinates. At each sampling period, the norm of $s = c_0 x_1 + x_2$, which measures the distance from the desired sliding surface $s = 0$, can be used to evaluate the one step performance. Therefore, we define the following utility function associated with the one-step cost at the ith sampling period,

$$U_i = U(s) \tag{2.4}$$

with

$$U(s) = \begin{cases} 0 & |s_1| < \delta_1, |s_2| < \delta_2, ..., |s_n| < \delta_n \\ 1 & \text{otherwise} \end{cases} \tag{2.5}$$

where s is defined in Equation (2.3) and $s = [s_1, s_2, ..., s_n]^T$, $|s_i|$ denotes the absolute value of the ith component of the vector s, the parameter $\delta_i > 0$ for $i = 1, 2, ..., n$. At each step, there is a value U_i and the total cost starting from the kth step along the infinite time horizon can be expressed as follows:

$$J_k = J(x(k), \overline{u}(k)) = \sum_{i=k}^{\infty} \gamma^{i-k} U_i, \tag{2.6}$$

where $x(k)$ is the state vector of system (2.1) sampled at the kth step with $x(k) = [x_1^T(k), x_2^T(k)]^T$, γ is the discount factor with $0 < \gamma < 1$, and $\overline{u}(k) = (u_k, u_{k+1}, ..., u_\infty)$ is the control sequence starting from the kth step. Note that for the deterministic system (2.1), the preceding states after the kth step are determined by $x(k)$ and the control sequence $\overline{u}_k$. Accordingly, J_k is a function of $x(k)$ and $\overline{u}(k)$ with $J_k = J(x(k), \overline{u}(k))$. Also note that both the cost function J_k and the utility function U_k are defined based on the discrete samplings of the continuous system (2.1). Now, we can define the problem of controlling the sensor–actuator system (2.1) in this framework as follows:

$$\min_{u(0), u(1), ..., u(\infty) \in \Omega} \quad J_0 = \sum_{i=0}^{\infty} \gamma^i U_i$$

subject to: $\tag{2.7a}$

$$\begin{cases} \dot{x}_1(t) = x_2(t) \\ \dot{x}_2(t) = -D^{-1}(t)(x_1(t))(u(t) + C(x_1(t), x_2(t))x_2(t) + \phi(x_1(t))) \end{cases} \tag{2.7b}$$

$$u(t) = u(i) \quad \text{for} \quad i\tau \le t < (i+1)\tau, \tag{2.7c}$$

where U_i is defined by Equations (2.4) and (2.5), $\tau > 0$ is the sampling period, the set Ω defines the feasible control actions, and J_0 is the cost function for $k = 0$ in Equation (2.6). It is worth noting that J_0 is a function of $\overline{u}(0) = (u_0, u_1, ..., u_\infty)$ and $x(0)$ according to Equation (2.6). The optimization in Equation (2.7) is relative to $\overline{u}(0)$ with a given initial state $x(0)$. Also note that in the optimization problem in Equation (2.7), the decision variables $u(0), u(1), ..., u(\infty)$ are defined in every sampling period. The control action keeps the value in the duration of two consecutive sampling steps. This formulation is consistent with the real implementations of digital controllers.

Remark 2.2 There are infinitely many decision variables, which are $u(0), u(1), ...,$ $u(\infty)$, in the optimization problem in Equation (2.7). Therefore, this is an infinite dimensional problem. It cannot be solved directly using numerical methods. Conventionally, such kind of problem is often solved by using a finite dimensional approximation. In addition, note that the dynamic model of the system appears in the optimization problem in Equation (2.7) and it will also show up in the finite dimensional relaxation of the problem, which means the resulting solution requires model information and thus is also model-dependent. In contrast, in this chapter we investigate the model-independent variable structure control of sensor–actuator systems on the infinite time horizon.

2.4 Model-Free Control of the Euler–Lagrange System

In this section, we present the strategy to solve the constrained optimization problem efficiently without knowing the model information of the chaotic system. We first investigate the optimality condition of Equation (2.7) and present an iterative procedure to

approach the analytical solution. Then, we analyze the convergence of the iterative procedure and the stability with the derived control strategy.

2.4.1 Optimality Condition

Denoting J^* the optimal value to the optimization problem in Equation (2.7), i.e.

$$J^* = \min_{u(0), u(1), \ldots, u(\infty) \in \Omega} J_0$$

subject to: (2.7b), (2.7c)　　　　　　　　　　　　　　　　　　(2.8)

According to the principle of optimality [23], the solution of Equation (2.7) satisfies the
following Bellman equation:

$$J^*(y) = \min_{u_k \in \Omega}(U_k + \gamma J^*(z)) \quad \forall x, \forall k = 0, 1, 2, \ldots \quad (2.9)$$

where z is the solution of Equation (2.7b) at $t = k + 1$ with $x(k) = y$ and the control
action $u(t) = u_k$ for $k\tau \leq t < (k + 1)\tau$. Without introducing confusion, we simply write
Equation (2.9) as follows:

$$J^* = \min(U_k + \gamma J^*). \quad (2.10)$$

Define the Bellman operator $\mathcal{B}$ relative to function $h(z)$ as follows:

$$\mathcal{B}h(z) = \min(U_k + \gamma h(z)). \quad (2.11)$$

Then, the optimality condition in Equation (2.10) can be simplified into the following
with the Bellman operator,

$$J^* = \mathcal{B}J^*. \quad (2.12)$$

Note that the function U_k is implicitly included in the Bellman operator. Equation (2.12)
constitutes the optimality condition for the problem in Equation (2.7). It is difficult to
solve the explicit form of J^* analytically from Equation (2.9). However, it is possible to
get the solution by iterations. We use the following iterations to solve J^*,

$$\hat{J}(n + 1) = \mathcal{B}\hat{J}(n)$$

subject to: (2.7b), (2.7c)　　　　　　　　　　　　　　　　　　(2.13)

The control action keeps constant in the duration between the kth and the k+1th step,
i.e. $u^*(t) = u_k^*$ for $k\tau \leq t < (k + 1)\tau$. u_k^* can be obtained from Equation (2.9) based on
Equation (2.13),

$$u_k^* = \mathrm{argmin}_{u_k \in \Omega}(U_k + \gamma J^*). \quad (2.14)$$

2.4.2 Approximating the Action Mapping and the Critic Mapping

In the previous sections, the iteration (2.13) is derived to calculate J^* and the optimization (2.14) is obtained to calculate the control law. The iteration to approach J^* and the
optimization to derive u^* have to be run in every time step in order to obtain the most
up-to-date values. Inspired by the learning strategies widely studied in artificial intelligence, a learning-based strategy is used in this section to facilitate the processing. After a
sufficiently long time, the system is able to memorize the mapping of J^* and the mapping
of u^*. After this learning period, there will be no need to repeat any iterations or optimal
searching, which will make the strategy more practical.

Note that the optimal cost J^* is a function of the initial state. Counting the cost from the current time step, J^* can also be regarded as a function of both the current state and the optimal action at the current time step according to Equation (2.10). Therefore, $\hat{J}(n)$, the approximation of J^*, can also be regarded as a function relative to the current state and the current optimal input. As to the optimal control action u^*, it is a function of the current state. Our goal in this section is to obtain the mapping from the current state and the current input to $\hat{J}(n)$ and the mapping from the current state to the optimal control action u^* using parameterized models, denoted as the critic model and the action model, respectively. Therefore, we can write the critic model and the action model as $J_n(u_n^*, x_n, W_c)$ and $u_n^*(x_n, W_a)$, respectively, where W_c and W_a are the parameters of the critic model and the action model, respectively.

In order to train the critic model with the desired input–output correspondence, we define the following error at time step $n + 1$ to evaluate the learning performance,

$$e_c(n + 1) = B\hat{J}(n) - \hat{J}(n + 1),$$

$$E_c(n + 1) = \frac{1}{2}e_c^2(n + 1). \tag{2.15}$$

Note that $B\hat{J}(n)$ is the desired value of $\hat{J}(n + 1)$ according to Equation (2.13). Using the back-propagation rule, we get the following rule for updating the weight W_c of the critic model,

$$W_c(n + 1) = W_c(n) + \delta W_c(n)$$

$$= W_c(n) - l_c(n)\frac{\partial E_c(n)}{\partial W_c(n)}$$

$$= W_c(n) - l_c(n)\frac{\partial E_c(n)}{\partial \hat{J}(n)}\frac{\partial \hat{J}(n)}{\partial W_c(n)}, \tag{2.16}$$

where $l_c(n)$ is the step size for the critic model at the time step n.

As to the action model, the optimal control u^* in Equation (2.14) is the one that minimizes the cost function. Note that the possible minimum cost is zero, which corresponds to the scenario with the state staying inside the desired bounded area. In this regard, we define the action error as follows:

$$e_a(n) = \hat{J}_n,$$

$$E_a(n) = \frac{1}{2}e_a^2(n). \tag{2.17}$$

Then, similar to the update rule of W_c for the critic model, we get the following update rule of W_a for the action model,

$$W_a(n + 1) = W_a(n) - l_a(n)\frac{\partial E_a(n)}{\partial \hat{J}(n)}\frac{\partial \hat{J}(n)}{\partial u(n)}\frac{\partial u(n)}{\partial W_a(n)}, \tag{2.18}$$

where $l_a(n)$ is the step size for the action model at the time step n.

Equations (2.16) and (2.18) update the critic model and the action model progressively. After W_c and W_a have learnt the model information by learning for a sufficiently long time, their values can be fixed at the one obtained at the final step and no further learning is required, which is in contrast to Equation (2.14) which requires an optimization problem to be solved even after a long time.

Figure 2.1 The cart–pole system.

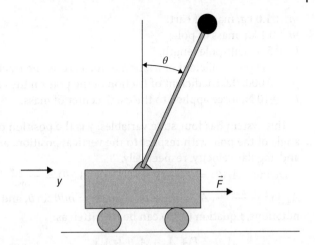

2.5 Simulation Experiment

In this section, we consider the simulation implementation of the proposed control strategy. The dynamics given in Equation (2.1) model a wide class of sensor–actuator systems. Particularly, to demonstrate the effectiveness of the proposed self-learning variable structure method, we apply it to the stabilizations of a typical benchmark system: the cart–pole system.

The cart–pole system, as sketched in Figure 2.1, is a widely used testbed for the effectiveness of control strategies. The system is composed of a pendulum and a cart. The pendulum has its mass above its pivot point, which is mounted on a cart moving horizontally. In this section, we apply the proposed control method to the cart–pole system to test the effectiveness of our method.

2.5.1 The Model

The cart–pole model used in this work is the same as that in [24], which can be described as follows:

$$\ddot{\theta} = \frac{g \sin\theta + \cos\theta[-F - ml\dot{\theta}^2 \sin\theta + \mu_c \text{sgn}(\dot{y})] - \frac{\mu_p \dot{\theta}}{ml}}{l\left(\frac{4}{3} - \frac{m\cos^2\theta}{m_c + m}\right)}, \tag{2.19}$$

$$\ddot{y} = \frac{F + ml[\dot{\theta}^2 \sin\theta - \ddot{\theta}\cos\theta] - \mu_c \text{sgn}(\dot{y})}{m_c + m}, \tag{2.20}$$

where

$$\text{sgn}(x) = \begin{cases} 1, & \text{if} \quad x > 0 \\ 0, & \text{if} \quad x = 0 \\ -1, & \text{if} \quad x < 0 \end{cases} \tag{2.21}$$

with the following values for the parameters:

g : 9.8 m/s^2, acceleration due to gravity;

m_c : 1.0 kg, mass of cart;
m : 0.1 kg, mass of pole;
l : 0.5 m, half-pole length;
μ_c : 0.0005, coefficient of friction of the cart on the track;
μ_p : 0.000002, coefficient of friction of the pole on the cart;
F : ±10 N, force applied to the cart's center of mass.

This system has four state variables: y is the position of the cart on the track, θ is the angle of the pole with respect to the vertical position, and $\dot{y}$ and $\dot{\theta}$ are the cart velocity and angular velocity, respectively.

Define $A_1(\theta) = -\frac{l}{\cos\theta}(\frac{4}{3} - \frac{m\cos^2\theta}{m_c+m})$, $A_2(\theta) = -\frac{g\sin\theta}{\cos\theta}$, $A_3(\theta,\dot{\theta}) = ml\dot{\theta}\sin\theta + \frac{\mu_p}{ml\cos\theta}$, $A_4(\dot{y}) = -\frac{\mu_c\text{sgn}(\dot{y})}{\dot{y}}$, $A_5 = m_c + m$, $A_6(\theta,\dot{\theta}) = ml\dot{\theta}\sin\theta$, and $A_7(\theta) = -ml\cos\theta$. With these notations, Equation (2.19) can be rewritten as:

$$A_1\ddot{\theta} = F + A_2 + A_3\dot{\theta} + A_4\dot{y}.$$

$$\frac{A_1A_5}{A_1+A_7}\ddot{y} = F + \frac{A_2A_7}{A_1+A_7} + \frac{A_1A_6+A_3A_7}{A_1+A_7}\dot{\theta} + \frac{A1A_4+A_4A_7}{A_1+A_7}\dot{y}. \tag{2.22}$$

By choosing

$$D = \begin{bmatrix} A_1 & 0 \\ 0 & \frac{A_1A_5}{A_1+A_7} \end{bmatrix}, C = -\begin{bmatrix} A_3 & A_4 \\ \frac{A_1A_6+A_3A_7}{A_1+A_7} & \frac{A_1A_4+A_4A_7}{A_1+A_7} \end{bmatrix}$$

$$\phi = -\begin{bmatrix} A_2 \\ \frac{A_2A_7}{A_1+A_7} \end{bmatrix}, q = \begin{bmatrix} \theta \\ y \end{bmatrix}, u = \begin{bmatrix} F \\ F \end{bmatrix}$$

the system of Equation (2.19) coincides with the model of Equation (2.1). Note that the input u in this situation is constrained in the set $\Omega = \{u = [u_1, u_2]^T, u_1 = u_2 \in \mathbb{R}\}$.

2.5.2 Experiment Setup and Results

In the simulation experiment, we set the discount factor $\gamma = 0.95$, the sliding surface parameter $k = 10$, $\delta_1 = 2$, and $\delta_2 = 24$. The feasible control action set Ω in Equation (2.7) is defined as $\Omega = \{u = [u_1, u_2]^T, u_1 \in \mathbb{R}, u_2 \in \mathbb{R}, u_1 = u_2 = \pm10\,\text{N}\}$. This definition corresponds to the widely used bang-bang control in industry. To make the output of the action model within the feasible set, the output of the action network is clamped to 10 if it is greater than or equal to zero and clamped to -10 if less than zero. The sampling period τ is set to 0.02 s. Both the critic model and the action model are linearly parameterized. The step size of the critic model, $l_c(n)$, and that of the action model, $l_a(n)$, are both set to 0.03. Both the update of the critic model weight W_c in Equation (2.16) and the update of the action model weight W_a in Equation (2.18) last for 30 s. For the uncontrolled cart–pole system with $F = 0$ in Equation (2.19), the pendulum will fall down. The control objective is to stabilize the pendulum to the inverted direction ($\theta = 0$). The time history of the state variables is plotted in Figure 2.2 for the system with the proposed self-learning variable structure control strategy. From Figure 2.2, it can be observed that θ is stabilized in a small vicinity around zero (with a small error of ±0.1 rad), which corresponds to the inverted direction.

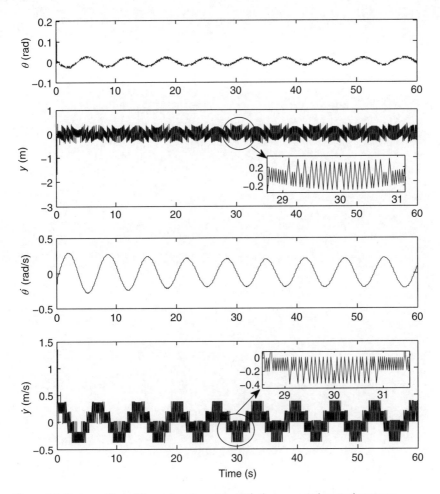

Figure 2.2 State profiles of the cart–pole system with the proposed control strategy.

2.6 Summary

In this chapter, the self-learning variable structure control is considered to solve a class of sensor–actuator systems. The control problem is formulated from the optimal control perspective and solved via iterative methods. In contrast to existing models, this method does not need pre-knowledge of the accurate mathematical model. The critic model and the the action model are introduced to make the method more practical. Simulations show that the control law obtained by the proposed method indeed achieves the control objective.

3

Projection Neural Networks for Robot Arm Control

3.1 Introduction

With advances in mechanics, electronics, and computer engineering, automatic manipulators are becoming increasingly popular in industrial applications to reduce the burden on labor forces. Among the various types of available manipulators, redundant manipulators, which usually possess more degrees of freedom than general manipulators and thus offer increased control flexibility for complicated tasks, have attracted intensive research in recent decades.

Despite the great advantages offered by redundant manipulators in dexterous manipulation for complicated tasks, the efficient control of such manipulators remains a challenging problem. A redundant manipulator provides a nonlinear mapping from its joint space to a Cartesian workspace. The goal of kinematic control is to find a control action in the joint space that produces a desired motion in the workspace. However, the nonlinearity of the mapping makes it difficult to directly solve this problem at the angle level. Instead, in most approaches, the problem is first converted into a problem at the velocity or acceleration level, and solutions are then sought in the converted space. In early work [25], control solutions were directly found using the pseudoinverse of the Jacobian matrix of a manipulator. However, this control strategy suffers from an intensive computational burden because of the need to perform the pseudoinversion of matrices continuously over time. It also suffers from local instability problems [26].

To overcome the problems with pseudoinverse-based solutions, later work [27] formulated the control problem as a quadratic programming problem to find an optimal solution with minimal kinematic energy consumption under physical constraints. Because these physical constraints typically include inequality constraints, the solution to such a problem usually cannot be obtained analytically, and serial processing techniques, e.g. matrix decomposition and Gaussian elimination, have been applied to these problems to obtain numerical solutions [28]. However, the relatively low efficiency of conventional serial processing techniques poses an additional hindrance to the real-time control of manipulators.

The parallel processing capability of neural networks has inspired researchers to explore the use of neural networks to control redundant manipulators. In [29], a neural network was constructed to learn the forward kinematics of a redundant manipulator and subsequently utilized to control such a system based on the learned model. In [30],

Kinematic Control of Redundant Robot Arms Using Neural Networks, First Edition.
Shuai Li, Long Jin and Mohammed Aquil Mirza.
© 2019 John Wiley & Sons Ltd. Published 2019 by John Wiley & Sons Ltd.

a feed-forward neural network was employed to learn the parametric uncertainties in the dynamical model of a robot manipulator. In [31], an approximately optimal resolution scheme was designed using an adaptive critic framework for the kinematic control of a redundant manipulator.

In addition to the use of feed-forward neural networks to control redundant manipulators, the application of recurrent neural networks (RNNs) for manipulator control has also received intensive research attention. In [32], the authors presented a Lagrange neural network model to minimize a cost function under an equation constraint. This model is able to address constrained quadratic programming problems by converting inequality constraints into equations using slack variables, and thus, this model is applicable for the kinematic control of redundant manipulators following its quadratic programming formulation. For the direct consideration of inequalities in constrained quadratic programming problems, researchers have attempted to consider the problem in its dual space. Various dual neural networks with different architectures have been presented. Typical works include [33, 35], in which analyses were conducted in the dual space and a convex projection function was often employed to represent inequality constraints. Methods of this type are highly efficient for real-time processing and have been successfully used in various applications, including k-winners-take-all problem solving [36] and image restoration [37]. Because of their generality in coping with constrained quadratic programming problems, they are potentially applicable for redundant manipulator control. In [38], for joint torque optimization, a unified quadratic-programming-based dynamical system approach was established for the control of redundant manipulators. The authors of [39] used a minimum-energy-based control objective to construct RNNs for the control of redundant manipulators. They also identified periodic oscillations in several neural networks when applied for manipulator control. A single-layered dual neural network was presented in [40] for the control of kinematically redundant manipulators. The presented neural structure offers reduced spatial complexity of the network and increased computational efficiency. An adaptive RNN with guaranteed asymptotic convergence was presented in [41] and successfully employed for the control of a mobile manipulator with unknown dynamics. In [42, 43], RNN approaches were extended to achieve multiple manipulator cooperation and coordination.

Although great success has been achieved in the kinematic control of redundant manipulators by using RNN approaches, the existing solutions still have certain unsatisfactory aspects that severely restrict the wide application of RNNs in industry for redundant manipulator control. This chapter identifies two particular limitations of existing RNN control solutions. The first limitation is that the position control error of existing solutions accumulates over time, which prohibits such solutions from being used in applications involving tasks that must run for a long time in the presence of additive noise. The second limitation is that in existing methods, the projection set is assumed to be convex, which eliminates many real projection operations that are utilized in industry from consideration. For instance, the bang-bang controllers that are commonly utilized in many practical control systems cannot be characterized in terms of any convex projection operators [44] and thus cannot be implemented using existing RNN methods. The possibility of modifying RNNs to allow nonconvex projection sets while guaranteeing stable manipulator control remains unexplored. This chapter makes progress on this front through the proposal of modified neural networks to overcome the two above mentioned limitations of existing neural approaches. To the best of our knowledge, this is the first neural network model for redundant manipulator control that is subject to neither position error accumulation nor the restriction that the projection set must be convex.

3.2   Problem Formulation

The forward kinematics of a manipulator involves a nonlinear transformation from a joint space to a Cartesian workspace, as described by

$$r(t) = f(\theta(t)), \tag{3.1}$$

where $r(t)$ is an m-dimensional vector in the workspace that describes the position and orientation of the end-effector at time t; $\theta(t)$ is an n-dimensional vector in the joint space, each element of which describes a joint angle; and $f(\cdot)$ is a nonlinear mapping determined by the structure and parameters of the manipulator. Because of the nonlinearity and redundancy of the mapping $f(\cdot)$, it is usually difficult to directly obtain the corresponding $\theta(t)$ for a desired $r(t) = r_d(t)$. By contrast, the mapping from the joint space to the workspace at the velocity level is an affine mapping and thus can be used to significantly simplify the problem, which can be illustrated as described in the following. Computing the time derivative on both sides of (3.1) yields

$$\dot{r}(t) = J\dot{\theta}(t), \tag{3.2}$$

where $J = \partial f/\partial\theta \in \mathbb{R}^{m\times n}$ is the Jacobian matrix of $f(\cdot)$, and $\dot{r}(t)$ and $\dot{\theta}(t)$ are the Cartesian velocity and the joint velocity, respectively. Usually, the motion control of a manipulator can be partitioned into two loops: one is an external loop for manipulator workspace tracking control; and the other is an internal loop for motor speed control. A practical servo motor with dedicated speed controllers is able to swiftly reach a reference speed. Under the condition that the time scale of the internal control loop is much shorter than that of the external loop, i.e. the internal loop reaches its reference speed much faster than the stabilization of the external loop control, the transition to a steady state in the internal loop can be neglected and the external loop control can be designed without direct consideration of the internal loop dynamics. In this chapter, we assume that the internal motor speed control loop is sufficiently fast in comparison with the external loop dynamics and we focus on the design of the external control loop. Each element of $\dot{\theta}(t)$, which is the angular speed of a particular joint of the manipulator, serves as an input to the external loop. Define an n-dimensional input vector $u(t) = \dot{\theta}(t)$. Then, the manipulator kinematics can be rewritten as follows:

$$\dot{r}(t) = Ju(t). \tag{3.3}$$

Because of the physical limitations of motors, the angular speed of each joint is limited to a certain bounded range. For example, the constraint $u(t) \in \Omega = [-\eta, \eta]^n$ applies for the case in which each individual joint is restricted to a maximum speed of η. To capture the restrictions on $u(t)$, we set the following general constraint:

$$u(t) \in \Omega, \tag{3.4}$$

where Ω is a set in an n-dimensional space. For a redundant manipulator described by (3.3) that is subject to the control input constraint (3.4), we wish to find a control law $u(t)$ such that the tracking error $e(t) = r(t) - r_d(t)$ for a given reference trajectory $r_d(t)$ converges over time.

3.3 A Modified Controller without Error Accumulation

In this section, we first examine an existing RNN designed for the control of redundant manipulators. Then, we modify this existing RNN to address the error accumulation problem for position control. Finally, we prove the stability of the presented controller using Lyapunov theory.

3.3.1 Existing RNN Solutions

As reviewed in Section 3.1, the kinematic control of redundant manipulators using RNNs has been extensively studied in recent decades. Although existing methods of this type differ in the objective functions or neural dynamics used, most of them follow similar design principles: the redundant manipulator control problem is typically formulated as a constrained quadratic optimization problem, which can be equivalently converted into a set of implicit equations. Then, a convergent RNN model, the equilibrium of which is identical to the solution of this implicit equation set, is devised to solve the problem recursively. Without loss of generality, the corresponding optimization problem, with the joint space kinematic energy $u^T u = \dot{\theta}^T \dot{\theta}$ as the objective function (where the superscript ^{T} denotes the transpose of a vector or matrix) and with a set constraint on the joint velocity $u \in \Omega$, can be presented as follows:

$$\min_u u^T u, \tag{3.5}$$

$$\dot{r}_d = Ju, \tag{3.6}$$

$$u \in \Omega, \tag{3.7}$$

where Ω is a set. In most of the existing literature [38–40, 45, 46], the set Ω is chosen to be $\Omega = \{x \in \mathbb{R}^n, -\eta^- \le x \le \eta^+\}$ to capture the physical constraints on the joint speeds. With the selection of a Lagrange function $L(u, \lambda) = u^T u/2 + \lambda^T (\dot{r}_d - Ju)$, where $\lambda \in \mathbb{R}^m$ is the Lagrange multiplier corresponding to the equality constraint (3.6), the optimal solution to (3.5)–(3.7) is equivalent to the solution of the following equation set, according to the so-called Karush–Kuhn–Tucker condition [11]:

$$u = P_\Omega \left(u - \frac{\partial L}{\partial u} \right), \tag{3.8a}$$

$$\dot{r}_d = Ju. \tag{3.8b}$$

A projected RNN for solving (3.8) can be designed as follows:

$$\varepsilon \dot{u} = -u + P_\Omega \left(u - \frac{\partial L}{\partial u} \right) = -u + P_\Omega(J^T \lambda), \tag{3.9a}$$

$$\varepsilon \dot{\lambda} = \dot{r}_d - Ju, \tag{3.9b}$$

where $\varepsilon > 0$ is a scaling factor and $P_\Omega(\cdot)$ is a projection operation to a set Ω, which is defined as $P_\Omega(x) = \mathrm{argmin}_{y \in \Omega} \| y - x \|$. The dynamics of (3.9) has been rigorously proven to converge to an equilibrium that is identical to the optimal solution of (3.5) [39, 40, 47].

The objective function in (3.5) can include an additional term $c_1(\dot{r}_d - Ju)^T(\dot{r}_d - Ju)$ to increase the penalty on the velocity-level tracking error $\dot{r}_d - Ju$; in this case, the overall optimization is reformulated as follows:

$$\min_u u^T u + c_1(\dot{r}_d - Ju)^T(\dot{r}_d - Ju), \tag{3.10}$$

$$\dot{r}_d = Ju, \tag{3.11}$$

$$u \in \Omega, \tag{3.12}$$

where $c_1 > 0$ is a weight. Note that the solution to (3.10)–(3.12) is identical to that of (3.5)–(3.7) because the additional term $c_1(\dot{r}_d - Ju)^T(\dot{r}_d - Ju)$ is equal to zero because of the requirement that $\dot{r}_d = Ju$ in the formulation. For (3.10)–(3.12), the Lagrange function is $L(u, \lambda) = u^T u/2 + c_1(\dot{r}_d - Ju)^T(\dot{r}_d - Ju)/2 + \lambda^T(\dot{r}_d - Ju)$, and a generalized projected RNN can be finally obtained as follows:

$$\varepsilon \dot{u} = -u + P_\Omega(J^T \lambda + c_1 J^T(\dot{r}_d - Ju)), \tag{3.13a}$$

$$\varepsilon \dot{\lambda} = \dot{r}_d - Ju. \tag{3.13b}$$

Although the RNN model expressed in (3.13) generalizes that of (3.9) by introducing an additional design degree of freedom, they both have identical equilibrium points, achieve redundancy resolution at the velocity level, and are subject to certain limitations, as discussed in detail in the next subsection. Without loss of generality, we proceed with our discussion based on the basic RNN model (3.9).

To summarize the discussion presented in this section, conventional RNN-based manipulator control involves three steps: (1) model the manipulator control problem as a constrained optimization problem [see Equations (3.5)–(3.7)]; (2) investigate the optimization problem in the dual space and find the expression of its solution in the form of a set of nonlinear equations [see Equation (3.9)]; and (3) devise a convergent neural dynamics whose steady-state solution is identical to that of the nonlinear equation set [see Equation (3.9)].

Note that the Jacobian matrix J in the optimization problem defined by (3.5)–(3.7) varies with time because of the movement of the manipulator. Accordingly, the optimal solution to this optimization problem, which is the desired control action, also varies with time. The continuously time-varying property of the Jacobian matrix implies that the desired control action also varies continuously with time. This fact further implies that there may be no need to recompute the control action every time; instead, it may be possible to evolve each real-time control action from its predecessor, thereby reducing the necessary computation. In other words, the historical data regarding the control actions can be leveraged for rapid computations for real-time control. For example, the RNN in (3.13) computes the difference between two consecutive control actions, thereby significantly reducing computational costs.

3.3.2 Limitations of Existing RNN Solutions

In this subsection, we discuss two limitations of existing RNN solutions: the accumulation of position control error in the workspace over time; and the fact that the projection operations considered in existing RNNs do not admit nonconvex sets.

We first show that (3.9) is subject to drift in the workspace in the tracking of r_d. Although it can be proven that $\dot{r} - \dot{r}_d = Ju - \dot{r}$ converges to zero, the error $e = r - r_d$ of (3.9) accumulates over time when input noise is considered. To illustrate this, we first express λ in terms of u according to (3.9b) as follows:

$$\lambda = \lambda_0 + \frac{1}{\varepsilon} \int_0^t (\dot{r}_d - Ju) dt$$

$$= \lambda_0 + \frac{1}{\varepsilon}(r_d - r_{d0}) - \frac{1}{\varepsilon} \int_0^t Ju \, dt, \tag{3.14}$$

where $\lambda_0 = \lambda(0)$ represents the value of λ at time $t = 0$, $\lambda = \lambda(t)$, and $r_{d0} = r_d(0)$ is the desired coordinate of the end-effector in the workspace at $t = 0$. Substituting (3.14) into (3.9a) yields

$$\varepsilon \dot{u} = -u + P_\Omega \left(J^T [\lambda_0 + \frac{1}{\varepsilon}(r_d - r_{d0}) - \frac{1}{\varepsilon} \int_0^t Ju \, dt] \right). \tag{3.15}$$

For u in (3.3), it is found that

$$\int_0^t Ju \, dt = r - r_0, \tag{3.16}$$

where $r_0 = r(0)$ is the workspace coordinate of the end-effector at time $t = 0$. Accordingly, the control input u in (3.15) can be rewritten as

$$\varepsilon \dot{u} = -u + P_\Omega \left(J^T \left[\lambda_0 + \frac{(r_d - r) - (r_{d0} - r_0)}{\varepsilon} \right] \right). \tag{3.17}$$

Regarding the control law expressed in (3.17), we offer the following remark.

Remark 3.1 The desired trajectory r_d in the workspace appears in (3.17) in the form of $(r_d - r) - (r_{d0} - r_0)$ rather than simply $r_d - r$. Note that the former expression penalizes the difference between the tracking error $r_d - r$ at time t and its initial value $r_{d0} - r_0$, instead of the tracking error itself. Consequently, any initial tracking error at time $t = 0$ remains as a constant bias in the tracking error at time $t > 0$. Because of the relativity of the selection of the zero-time point for a time-invariant system, we conclude that any tracking error impacts all subsequent tracking errors, resulting in the accumulation of tracking error in the workspace.

Additionally, regarding the set Ω in (3.4), we offer the following remark.

Remark 3.2 To the best of our knowledge, all existing results on manipulator control using projected neural networks assume that the set Ω for projection is convex. This assumption excludes nonconvex sets from consideration. General conclusions regarding the relaxation of the convex constraint on Ω remain unexplored.

To summarize the above remarks, we identify the following limitations of existing RNN solutions for the kinematic control of redundant manipulators, which we wish to address in this chapter:

1) **Error Accumulation:** The position tracking error $e = r - r_d$ in the kinematic control of a redundant manipulator described by (3.3) using the control law (3.9) accumulates over time.
2) **Convexity Restriction:** In real applications, the set Ω may not be convex. With a nonconvex Ω, the control law expressed in (3.3) may lead to instability of a redundant manipulator.

In the following subsection, we propose a modified control law based on (3.17) to overcome the above limitations of the existing solutions.

3.3.3 The Presented Algorithm

To maintain the effectiveness of (3.9) for manipulator control, in this section, we modify the control law by retaining the negative feedback in (3.9) for control stability while introducing new elements to overcome the two limitations identified in Section 3.3.2.

To overcome the error accumulation limitation regarding position control, we first remove the terms $r_{d0} - r_0$ and λ_0 from the control law expressed in (3.17), resulting in

$$\varepsilon \dot{u} = -u + P_\Omega \left(\frac{1}{\varepsilon} J^{\mathrm{T}} (r_d - r) \right). \tag{3.18}$$

To avoid the transient violation of the constraint $u \in \Omega$, we simply replace (3.18) with its steady-state value, which satisfies this constraint. As a result, the presented control law is expressed as follows:

$$u = P_\Omega \left(\frac{1}{\varepsilon} J^{\mathrm{T}} (r_d - r) \right). \tag{3.19}$$

Equation (3.18) can be regarded as an alternative form of the presented control law (3.19) obtained by passing it through a first-order low-pass filter $\varepsilon \dot{u} = -u + y$, where $y = P_\Omega (J^{\mathrm{T}} (r_d - r)/\varepsilon)$. The low-pass filtering property of the control law (3.18) allows it to generate smoother control actions than those produced by (3.19). This property is useful in the practical implementation of the presented control law to avoid sharp changes in the control actions. However, the set constraint $u \in \Omega$ cannot be guaranteed to hold in transient states for (3.18). By contrast, $u \in \Omega$ holds unconditionally for (3.19).

With regard to the two limitations discussed in Section 3.3.2, we offer the following remark.

Remark 3.3 The presented control law (3.19) directly abides by the constraint $u \in \Omega$ at all times. As will be proven in the next subsection, the coordinate r of the end-effector in the workspace globally stabilizes to r_d for the presented control law, demonstrating that this control law remedies the problem of error accumulation in existing RNN-based solutions.

From the perspective of neural networks, we offer the following remark.

Remark 3.4 The control law expressed in (3.19) can be regarded as a single-layered feed-forward neural network with the position tracking error $r_d - r$ as the input and the control action u as the output. The weighting matrix of this neural network is J^T/ε, and the nonlinear activation function is $P_\Omega(\cdot)$.

3.3.4 Stability

In this section, we present a theorem and a theoretical proof regarding the stability of the presented control law (3.19).

Theorem 3.1 The control error $r - r_d$ of the control law expressed in (3.19) for a redundant manipulator described by (3.3) globally converges to zero, provided that the desired value r_d is a constant and that $0 \in \text{int}(\Omega)$, where $\text{int}(\Omega)$ denotes the interior of the set Ω.

Proof: The overall system dynamics can be expressed as follows by combining (3.3) and (3.19):

$$\dot{r} = JP_\Omega\left(\frac{1}{\varepsilon}J^T(r_d - r)\right). \tag{3.20}$$

For a constant r_d, its time derivative is always equal to zero, i.e. $\dot{r}_d = 0$. Accordingly, the dynamics of the error $e = r - r_d$ can be described by

$$\dot{e} = JP_\Omega\left(-\frac{1}{\varepsilon}J^T e\right). \tag{3.21}$$

Define a Lyapunov function $V = e^T e/2$. Its time derivative, given the dynamics of e, is

$$\dot{V} = e^T \dot{e} = e^T J P_\Omega\left(-\frac{1}{\varepsilon}J^T e\right). \tag{3.22}$$

According to the definition of the projection operation as $P_\Omega(z) = \text{argmin}_{y \in \Omega} \| y - z \|$, we have $\| P_\Omega(z) - z \| \leq \| y - z \|$, $\forall y \in \Omega$. Recall that $0 \in \text{int}(\Omega) \subseteq \Omega$. Choosing $y = 0$ and $z = -(1/\varepsilon)J^T e$ yields

$$\left\| P_\Omega\left(-\frac{1}{\varepsilon}J^T e\right) + \frac{1}{\varepsilon}J^T e \right\|^2 \leq \left\| \frac{1}{\varepsilon}J^T e \right\|^2. \tag{3.23}$$

Note that the left-hand side of the above expression can be written as

$$\left\| P_\Omega\left(-\frac{1}{\varepsilon}J^T e\right) + \frac{1}{\varepsilon}J^T e \right\|^2 = \left\| P_\Omega\left(-\frac{1}{\varepsilon}J^T e\right) \right\|^2 + \left\| \frac{1}{\varepsilon}J^T e \right\|^2$$
$$+ 2\left(\frac{1}{\varepsilon}J^T e\right)^T P_\Omega\left(-\frac{1}{\varepsilon}J^T e\right). \tag{3.24}$$

The expansion of (3.23) thus becomes

$$\left\| P_\Omega\left(-\frac{1}{\varepsilon}J^T e\right) \right\|^2 \leq -2\left(\frac{1}{\varepsilon}J^T e\right)^T P_\Omega\left(-\frac{1}{\varepsilon}J^T e\right), \tag{3.25}$$

which leads to

$$e^T J P_\Omega\left(-\frac{1}{\varepsilon}J^T e\right) \leq 0. \tag{3.26}$$

Therefore, we have

$$\dot{V} \leq 0. \tag{3.27}$$

We use LaSalle's invariant set principle [48] to find the largest invariant set. Letting $\dot{V} = 0$ yields

$$e^T J P_\Omega \left(-\frac{1}{\varepsilon} J^T e \right) = 0. \tag{3.28}$$

By considering this expression together with (3.25), it can be concluded that

$$\left\| P_\Omega \left(-\frac{1}{\varepsilon} J^T e \right) \right\|^2 \leq 0, \tag{3.29}$$

which implies that

$$P_\Omega \left(-\frac{1}{\varepsilon} J^T e \right) = 0. \tag{3.30}$$

According to the definition of the projection operator, (3.30) is equivalent to

$$\underset{y \in \Omega}{\text{argmin}} \left\| \frac{1}{\varepsilon} J^T e + y \right\| = 0. \tag{3.31}$$

Therefore,

$$\left\| \frac{1}{\varepsilon} J^T e \right\|^2 \leq \left\| \frac{1}{\varepsilon} J^T e + y \right\|^2, \quad \forall y \in \Omega, \tag{3.32}$$

i.e.

$$\| y \|^2 + \frac{2}{\varepsilon} y^T J^T e \geq 0, \quad \forall y \in \Omega. \tag{3.33}$$

Because $0 \in \text{int}(\Omega)$, there exists a $\rho > 0$ such that $z \in \text{int}(\Omega) \; \forall \| z \| \leq \rho$. Clearly, $z = -(\rho J^T e)/(2 \| J^T e \|)$ satisfies $\| z \| \leq \rho$ and thus is within the interior of the set Ω, i.e. $-(\rho J^T e)/(2 \| J^T e \|) \in \text{int}(\Omega) \subseteq \Omega$. Note that it is required that (3.33) must hold for all $y \in \Omega$, which includes $y = -(\rho J^T e)/(2 \| J^T e \|) \in \Omega$ as a special case. That is,

$$\left\| \frac{\rho J^T e}{2 \| J^T e \|} \right\|^2 - 2 \left(\frac{\rho J^T e}{2 \| J^T e \|} \right)^T \frac{1}{\varepsilon} J^T e \geq 0, \tag{3.34}$$

which is equivalent to

$$\frac{\rho \varepsilon}{4} \geq \| J^T e \| . \tag{3.35}$$

Recall that the value of ρ can be chosen to be as small as possible and that expression (3.35) holds for any $\rho > 0$. This implies that

$$\| J^T e \| = 0, \tag{3.36}$$

and, consequently,

$$J^T e = 0. \tag{3.37}$$

The Jacobian matrix $J \in \mathbb{R}^{m \times n}$ for a redundant manipulator satisfies $m < n$. For rank $(J) = m$, the solution of $e \in \mathbb{R}^m$ is unique and is $e = 0$. That is, the largest invariant set for $\dot{V} = 0$ contains only a single point, $e = 0$. We thus conclude that the error e of the manipulator with the presented control law exhibits global stability. This completes the proof. ∎

Regarding the requirement of $0 \in \text{int}(\Omega)$ in Theorem 3.1, we offer the following remark.

Remark 3.5   The requirement of $0 \in \text{int}(\Omega)$ in Theorem 3.1 can be separated into two parts: $0 \in \Omega$; and $0 \notin \partial\Omega$, where $\partial\Omega$ represents the boundary set of Ω. $0 \in \Omega$ is required to ensure the feasibility of the control action when the control objective is reached. When the control error is equal to zero, i.e. $r - r_d = 0$, the control action in (3.19) becomes $u = P_\Omega(0)$. The condition $0 \in \Omega$ guarantees that $u = P_\Omega(0)$ is well defined in this case. Notably, the condition on Ω in Theorem 3.1 is $0 \in \text{int}(\Omega)$, which is mild, allowing Ω to be nonconvex; to the best of our knowledge, this stands in contrast to all previous results for existing projected neural networks [49–56].

Regarding the computational efficiency of the presented control law (3.19), we offer the following remark.

Remark 3.6   In comparison with conventional non-neural approaches, which rely on pseudoinversion to compute control actions, e.g. $u = J^+ \dot{r}_d$, where $J^+ = J^T(JJ^T)^{-1}$ is the pseudoinverse of J, the presented control law is more efficient to compute. Note that pseudoinverse-based methods usually cannot abide by the set constraint; therefore, we set $\Omega = \mathbb{R}^n$ for a fair comparison. Moreover, recall that the multiplication of two matrices, one with dimensions of $l_1 \times l_2$ and the other with dimensions of $l_2 \times l_3$, requires $l_1 l_3(2l_2 - 1)$ flops of operations; the summation of two matrices with dimensions of $l_1 \times l_2$ requires $l_1 l_2$ flops of operations; and the inversion of a square matrix with dimensions of $l_1 \times l_1$ requires l_1^3 flops of operations [11]. The matrix J has dimensions of $m \times n$; therefore, $m^2(2n - 1)$ flops of operations are required to compute JJ^T, an additional m^3 flops are required to compute $(JJ^T)^{-1}$, and overall, $m^2(2n - 1) + m^3 + (mn(2m - 1)) = 4m^2 n + m^3 - m^2 - mn$ flops are required to complete the computation of J^+. In total, computing $u = J^+ \dot{r}_d$ requires $4m^2 n + m^3 - m^2 - mn + n(2m - 1) = 4m^2 n + m^3 - m^2 + mn - n$ flops. We can similarly analyze the time consumption of the presented control law $u = P_\Omega(J^T(r_d - r)/\varepsilon) = J^T(r_d - r)/\varepsilon$ when $\Omega = \mathbb{R}^n$. It can be found that the total number of flops required to compute a control action using the presented law is $2m^2 n - mn + m + n$. In the simulations of a PUMA 560 robot arm for tracking that are presented in this chapter, $m = 3$ and $n = 6$. In this case, the total numbers of flops are 246 and 99 for a pseudoinverse-based method and the proposed control law, respectively, from which we can see the improvement in computational efficiency offered by the presented controller.

3.4   Performance Improvement Using Velocity Compensation

In the previous section, we considered the regulation of r to a constant desired value r_d in the workspace for the kinematic control of redundant manipulators. In this section, we extend the presented algorithm to dynamic tracking by applying velocity compensation.

3.4.1   A Control Law with Velocity Compensation

As derived in the proof of Theorem 3.1, the overall dynamics of the system when the control law expressed in (3.19) is used can be written as

$$\dot{r} = J P_\Omega \left(\frac{1}{\varepsilon} J^T(r_d - r) \right). \tag{3.38}$$

To illustrate the limitations of (3.19) when it is used to track a dynamic signal r_d that is not constant with time, we consider the case in which $e = r - r_d = 0$. In this case, the right-hand side of (3.38) is zero, which cannot support the variation of r to follow the changes in r_d with time. To address this issue, we introduce an additional term into (3.19) as follows:

$$u = P_\Omega \left(\frac{1}{\varepsilon} J^T (r_d - r) + v \right), \tag{3.39}$$

where $v \in \mathbb{R}^n$ is used to compensate to satisfy the velocity requirement. To ensure that the desired velocity can also be reached in the steady state, we impose the requirement that $\dot{r}_d = Jv$. Because of the redundancy of the manipulator, the solution for such a v is not unique. We choose the solution with the minimum consumption of kinematic energy, i.e. the v that minimizes $v^T v$. The solution for v can thus be readily obtained as $J^+ \dot{r}_d$, where $J^+ = J^T (JJ^T)^{-1}$ is the pseudoinverse of J. Note that the expression for J^+ involves an inverse operation, which is computationally intensive. To avoid the direct computation of the inverse of a matrix, we design the following dynamics to recursively approach $v = J^+ \dot{r}_d$:

$$\zeta \dot{w} = \dot{r}_d - JJ^T w,$$
$$v = J^T w. \tag{3.40}$$

where $\zeta > 0$ is a scaling factor and $w \in \mathbb{R}^m$ is a co-state. By considering this dynamics together with (3.39), the proposed control law with velocity compensation can be written as follows:

$$u = P_\Omega \left(\frac{1}{\varepsilon} J^T (r_d - r) + J^T w \right), \tag{3.41a}$$

$$\zeta \dot{w} = \dot{r}_d - JJ^T w. \tag{3.41b}$$

Regarding the control law expressed in (3.41), we offer the following remark.

Remark 3.7   The control law expressed in (3.41) consists of two parts: a static feedback part, corresponding to the term $J^T (r_d - r)/\varepsilon$, to regulate the difference between r and r_d for position control; and a dynamic part, corresponding to $J^T w$, to regulate the difference between $\dot{r}$ and $\dot{r}_d$ for velocity compensation.

From the perspective of neural networks, we offer the following remark.

Remark 3.8   The control law expressed in (3.41) can be regarded as a recurrent neural network with r_d and r as the input, w as the hidden recurrent state, and u as the output. The mapping $P_\Omega(\cdot)$ serves as a nonlinear function in the output layer.

3.4.2 Stability

In this subsection, we present a stability analysis of the control law expressed in (3.41) for the case in which the desired workspace coordinate is varying with time.

Theorem 3.2   The control error $r - r_d$ of the control law expressed in (3.41) for a redundant manipulator described by (3.3) globally converges to zero, provided that $J^+ \dot{r}_d \in \text{int}(\Omega)$.

Proof: This proof is composed of two steps: (1) proof of the stability of the estimation, i.e. the dynamics of w in (3.41); and (2) proof of the stability of the tracking error, i.e. the dynamics of $e = r - r_d$.

Step 1: Proof of the stability of w in (3.41).

Consider $V_1 = (JJ^T w - \dot{r}_d)^T (JJ^T w - \dot{r}_d)/2$. The time derivative of V_1 given the dynamics of w can be written as

$$
\begin{aligned}
\dot{V}_1 &= (JJ^T w - \dot{r}_d)^T JJ^T \dot{w} \\
&= -\frac{1}{\zeta}(JJ^T w - \dot{r}_d)^T JJ^T (JJ^T w - \dot{r}_d) \\
&\leq -\frac{\lambda_0(JJ^T)}{\zeta}(JJ^T w - \dot{r}_d)^T (JJ^T w - \dot{r}_d) \\
&= -\frac{2\lambda_0(JJ^T)}{\zeta} V_1,
\end{aligned}
\tag{3.42}
$$

where $\lambda_1(JJ^T) > 0$ represents the smallest eigenvalue of JJ^T for $\text{rank}(JJ^T) = m$. It can be concluded from (3.42) that $JJ^T w - \dot{r}_d$ converges to zero at an exponential rate with a factor of $(2\lambda_0(JJ^T))/\zeta$. Therefore, w exponentially converges to $(JJ^T)^{-1}\dot{r}_d$.

Step 2: Proof of the stability of the tracking error $e = r - r_d$.

Because we have proven the exponential convergence of w to $(JJ^T)^{-1}\dot{r}_d$ in step 1, we consider the dynamics of e in the invariant set $w = (JJ^T)^{-1}\dot{r}_d$. With this, the control input u in (3.41) can be written as

$$
\begin{aligned}
u &= P_\Omega \left(\frac{1}{\varepsilon}J^T(r_d - r) + J^T(JJ^T)^{-1}\dot{r}_d \right) \\
&= P_\Omega \left(\frac{1}{\varepsilon}J^T(r_d - r) + J^+\dot{r}_d \right).
\end{aligned}
\tag{3.43}
$$

By considering this expression together with the manipulator kinematics, we obtain $\dot{r} = Ju = JP_\Omega(J^T(r_d - r)/\varepsilon + J^+\dot{r}_d)$. The tracking error dynamics for $e = r - r_d$ can thus be derived as follows:

$$
\dot{e} = JP_\Omega \left(-\frac{1}{\varepsilon}J^T e + J^+\dot{r}_d \right) - \dot{r}_d.
\tag{3.44}
$$

Note that $\dot{r}_d = JJ^T(JJ^T)^{-1}\dot{r}_d = JJ^+\dot{r}_d$. Accordingly,

$$
\dot{e} = J \left(P_\Omega \left(-\frac{1}{\varepsilon}J^T e + J^+\dot{r}_d \right) - J^+\dot{r}_d \right).
\tag{3.45}
$$

We define $V_2 = e^T e/2$ and compute its time derivative as

$$
\begin{aligned}
\dot{V}_2 &= e^T \dot{e} = e^T J \left(P_\Omega \left(-\frac{1}{\varepsilon}J^T e + J^+\dot{r}_d \right) - J^+\dot{r}_d \right) \\
&= -\varepsilon \left(\left(-\frac{1}{\varepsilon}J^T e + J^+\dot{r}_d \right) - J^+\dot{r}_d \right)^T \\
&\quad \cdot \left(P_\Omega \left(-\frac{1}{\varepsilon}J^T e + J^+\dot{r}_d \right) - J^+\dot{r}_d \right).
\end{aligned}
\tag{3.46}
$$

According to the definition of the projection operator, we have

$$
\left\| -\frac{1}{\varepsilon}J^T e + J^+\dot{r}_d - P_\Omega(-\frac{1}{\varepsilon}J^T e + J^+\dot{r}_d) \right\|^2 \leq \left\| -\frac{1}{\varepsilon}J^T e + J^+\dot{r}_d - J^+\dot{r}_d \right\|^2,
\tag{3.47}
$$

because $J^+\dot{r}_d \in \text{int}(\Omega) \subseteq \Omega$. Equivalently,

$$
\left\| P_\Omega(-\frac{1}{\varepsilon}J^T e + J^+\dot{r}_d) - J^+\dot{r}_d \right\|^2
$$

$$
\leq 2\left(\left(-\frac{1}{\varepsilon}J^T e + J^+\dot{r}_d - J^+\dot{r}_d\right)^T \left(P_\Omega\left(-\frac{1}{\varepsilon}J^T e + J^+\dot{r}_d\right) - J^+\dot{r}_d\right), \right. \tag{3.48}
$$

which, for $\dot{V}_2$ in (3.46), yields

$$
\dot{V}_2 \leq -\frac{\varepsilon}{2} \| P_\Omega(-\frac{1}{\varepsilon}J^T e + J^+\dot{r}_d) - J^+\dot{r}_d \|^2 \leq 0. \tag{3.49}
$$

To find the largest invariant set, let $\dot{V}_2 = 0$, and subsequently,

$$
P_\Omega(-\frac{1}{\varepsilon}J^T e + J^+\dot{r}_d) = J^+\dot{r}_d. \tag{3.50}
$$

Based on reasoning similar to that used in the proof of Theorem 3.1, the condition $J^+\dot{r}_d \in \text{int}(\Omega)$ and Equation (3.50) together imply that

$$
-\frac{1}{\varepsilon}J^T e + J^+\dot{r}_d = J^+\dot{r}_d \Rightarrow J^T e = 0. \tag{3.51}
$$

For rank$(J) = m$, the solution of $J^T e = 0$ for $e \in \mathbb{R}^m$ is unique and is $e = 0$. According to LaSalle's invariant set principle [48], we conclude that the system exhibits global stability with e converging to zero. This completes the proof. ∎

Remark 3.9 For a non-constant r_d, a steady-state input in (3.3) is required for tracking purposes. Because of the projection effect, a minimal requirement is that the set Ω must be sufficiently large to provide one feasible solution $u \in \Omega$ to support the desired velocity $\dot{r}_d$ in the workspace. Let us consider the case in which $\| \dot{r}_d \|$ is very large but the radius of the set Ω is very small. Clearly, there may not exist a value of $u \in \Omega$ such that $\dot{r}_d = Ju$. The condition that $J^+\dot{r}_d$ belongs to the set Ω in Theorem 3.2 is a quantitative description of such a requirement.

Regarding the computational efficiency of the presented control law (3.41), we offer the following remark.

Remark 3.10 Unlike conventional pseudoinverse-based methods, the control law expressed in (3.41) does not require the computationally intensive operation of matrix inversion. For the case of an update period of T and the set $\Omega = \mathbb{R}^n$, in discrete time, the control law becomes $u = J^T(r_d - r)/\varepsilon + J^T w$, $w = T(\dot{r}_d - JJ^T w^-)/\zeta + w^-$, where w^- denotes the value of w in the last time period before the update. Regarding the update of w, the computation of $J^T w^-$ requires $n(2m - 1)$ flops, the computation of $JJ^T w^-$ requires $m(2n - 1)$ additional flops, and overall, $4mn + 2m - n$ flops are required to complete the update of w. With the updated w, the computation of u requires $4mn + m$ additional flops. Overall, the total number of flops required to compute the control action in (3.41) is $8mn + 3m - n$. For the case of $m = 3$ and $n = 6$, as considered in the simulation-based verification study using a PUMA 560 robot arm, the total number of flops required is 147, which is less than the 246 flops required for pseudoinverse-based methods. From this difference in values, it is clear that the presented controller described by (3.41) demonstrates a higher computational efficiency than that of conventional pseudoinverse-based controllers.

Table 3.1 Summary of the Denavit–Hartenberg parameters of the PUMA 560 manipulator used in the simulation.

Link	a (m)	α (rad)	d (m)
1	0	$\pi/2$	0.67
2	0.4318	0	0
3	0.4318	$-\pi/2$	0.15005
4	0	$\pi/2$	0
5	0	$-\pi/2$	0
6	0	0	0.2

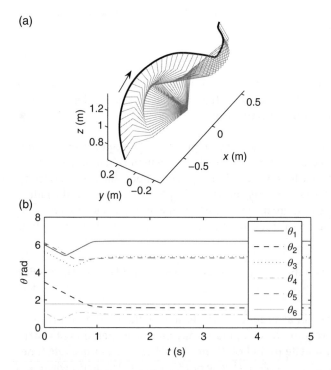

Figure 3.1 Simulation results for the position regulation control of the end-effector of a PUMA 560 to maintain a fixed position of [0.5500, 0, 1.3000] m in the workspace. (a) The end-effector trajectory and (b) the time history of the joint angle θ.

Additionally, regarding the impact of the internal loop dynamics, we offer the following remark.

Remark 3.11 In this chapter, our main results concern the external motion control loop for a manipulator. In a real implementation, the input signal $u(t)$ cannot be directly received by the manipulator and must be realized by means of the internal control loop. The effectiveness of the presented control method relies on the internal loop dynamics exhibiting much faster convergence than the external loop dynamics; therefore, the presented method imposes a corresponding requirement on the motor speed controller used in the inner loop.

3.5 Simulations

In this section, we consider numerical simulations of a PUMA 560 manipulator with the parameters summarized in Table 3.1 to demonstrate the effectiveness of the presented algorithms. The PUMA 560 is an articulated manipulator with six independently controlled joints. Its end-effector can reach any position in its workspace at any orientation. In these simulations, we considered only the three-dimensional position control of the end-effector, and thus, the PUMA 560 served as a redundant manipulator for this particular task.

3.5.1 Regulation to a Fixed Position

We first performed simulations to verify Theorem 3.1 by using the control law expressed in (3.19) to drive the manipulator's end-effector to a fixed position of $[0.5500, 0, 1.3000]$ m in Cartesian space. For the scaling coefficient $\varepsilon = 0.01$ and the set $\Omega = [-2, 2]^6$, a typical simulation run generated with a random initialization is shown in Figure 3.1.

(a)

(b)

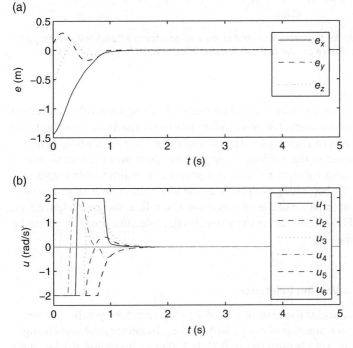

Figure 3.2 Simulation results for the position regulation control of the end-effector of a PUMA 560 to maintain a fixed position of $[0.5500, 0, 1.3000]$ m in the workspace. Time history of (a) the control error e and (b) the control action u.

(a)

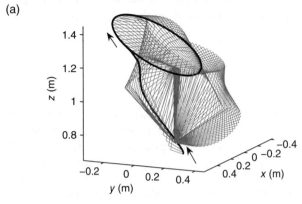

(b)

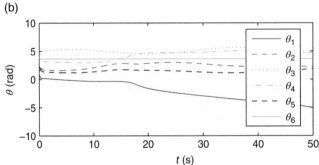

Figure 3.3 Simulation results for the tracking control of the end-effector of a PUMA 560 with respect to a time-varying reference along a circular path. (a) The end-effector trajectory and (b) the time history of the joint angle θ.

After a short transient state, the joint angle θ converges to a constant value, as shown in Figure 3.1a, and correspondingly, the end-effector position r reaches a constant value in the workspace, as shown in Figure 3.1b. The control error $e = r - r_{\mathrm{d}}$, where r_{d} represents the reference position in the workspace, approaches zero over time, as shown in Figure 3.2a. Notably, the control input u shown in Figure 3.2b remains within the set Ω at all times. It saturates at the beginning of the simulation when the error is large and converges to zero with the attenuation of the control error. Overall, as shown in Figure 3.1a, the end-effector of the PUMA 560 successfully reaches the reference position under the simulated control scheme.

3.5.2 Tracking of Time-Varying References

In this subsection, we consider the case of a time-varying reference position to verify Theorem 3.2. The reference position of the PUMA 560's end-effector moves at an angular speed of 0.2 rad/s along a circle centered at [0.25, 0, 1.3] m with a radius of 0.3 m and a revolution angle of 30° around the x axis. For a control action scaling factor of $\varepsilon = 0.01$, a co-state scaling factor of $\zeta = 0.001$, and $\Omega = [-2, 2]^6$, a typica simulation run using the control law expressed in (3.41) is presented in Figures 3.3 and 3.4. The joint angle

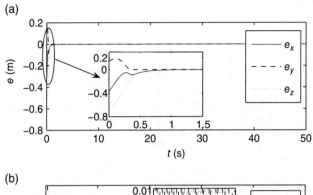

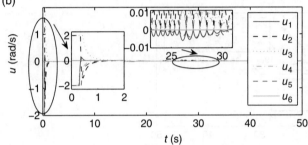

Figure 3.4 Simulation results for the tracking control of the end-effector of a PUMA 560 with respect to a time-varying reference along a circular path. Time history of (a) the control error e and (b) the control action u.

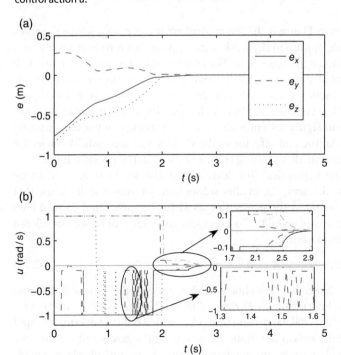

Figure 3.5 Simulation results obtained using the presented control laws with a nonconvex projection set. Time history of (a) the control error e with (3.19) and (b) the control action u with (3.19).

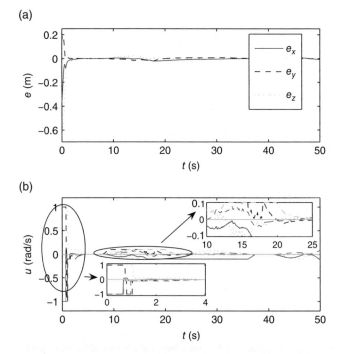

Figure 3.6 Simulation results obtained using the presented control laws with a nonconvex projection set. Time history of (a) the control error e with (3.41) and (b) the control action u with (3.41).

θ varies with time, as seen in Figure 3.3b. The control error $e = r - r_{\mathrm{d}}$, where r_{d} represents the circular reference motion in the workspace, approaches zero over time after a short transient state, as shown in Figure 3.4a. The control input u shown in Figure 3.4b remains within the set Ω at all times, with a saturation at the beginning when the error is large. Notably, the control action u does not converge to zero, as is required to compensate for the variation in the reference position with time. The fluctuations in u shown in the second inset figure in Figure 3.4b provide this time-varying compensation. Overall, as shown in Figure 3.3a, the end-effector of the PUMA 560 successfully tracks the time-varying reference motion after starting from a random initial configuration.

As a supplement to the theoretical justification that the set Ω is allowed to be non-convex in Theorems 3.1 and 3.2, in this subsection, we numerically show that nonconvex sets can be chosen as the projection set Ω for the presented control laws (3.19) and (3.41) without loss of stability. To exemplify the choice of Ω, we considered the following set in the simulation:

$$\Omega = \{x = [x_i] \in \mathbb{R}^6, -c_2 \le x_i \le c_2, \text{or } x_i = \pm c_1\}, \tag{3.52}$$

with $c_1 = 1$ and $c_2 = 0.1$. Note that this choice of Ω is nonconvex because $\mathbf{0} \in \Omega$ and $\mathbf{1} \in \Omega$ but $(\mathbf{0} + \mathbf{1})/2 \notin \Omega$. Physically, the Ω defined in (3.52) is generalized from commonly used strategies in industrial bang-bang control, in which only the maximum input action c_1, the negative of the maximum input action $-c_1$, and a zero input action 0 are valid. To avoid chattering phenomena in conventional bang-bang control, it is preferable to expand the zero input action to a small range $[-c_2, c_2]$, thus leading to the definition

Table 3.2 Comparisons of different RNN algorithms for the tracking control of a PUMA 560 manipulator.

	Non-convex Ω	$u(t) \in \Omega\Delta t$	Initial position	Number of neurons	Convergence	Regulation error	Tracking error	Error accumulation	Acceleration vs. velocity
Controller (3.19)	Yes	Yes	Any	6	Yes	Zero	Non-zero	No	Velocity
Controller (3.41)	Yes	Yes	Any	9	Yes	Zero	Zero	No	Velocity
Controller (3.9) [40]	No	No	Restrictive[a]	9	Yes	Fail[b]	Zero	Yes	Velocity
Controller in [39]	No	No	Restrictive[a]	15	Yes	Fail[b]	Zero	Yes	Velocity
Controller in [38]	No	No	Restrictive[a]	6	Yes	Fail[b]	Zero	Yes	Acceleration
Controller in [45]	No	No	Restrictive[a]	9	Yes	Fail[b]	Zero	Yes	Acceleration

a) For the controllers presented in [38–40, 45], the end-effector's initial position is required to be on the desired trajectory for tracking.
b) The controllers presented in [38–40, 45] are able to achieve the tracking of a time-varying reference position but fail for position regulation to a fixed reference position.

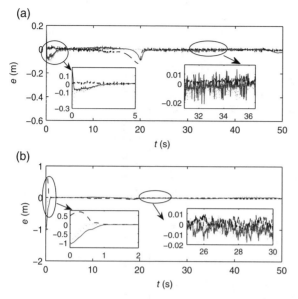

Figure 3.7 Simulation comparisons for the tracking control of the end-effector of a PUMA 560 with respect to a time-varying reference along a circular path using different neuro-controllers, i.e. controller 1 (control law (3.19) presented in this chapter), controller 2 (control law (3.41) presented in this chapter), controller 3 [40], and controller 4 [39], in the presence of random Gaussian noise at different levels of $\sigma = 0.01, 0.1, 1$. (a) Controller 1, noise level $\sigma = 0.01$ and (b) controller 1, noise level $\sigma = 0.1$.

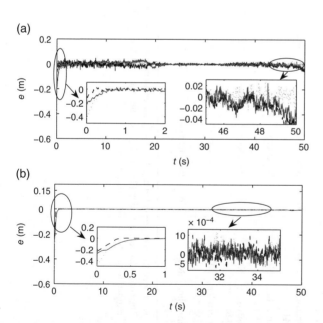

Figure 3.8 Simulation comparisons for the tracking control of the end-effector of a PUMA 560 with respect to a time-varying reference along a circular path using different neuro-controllers, i.e. controller 1 (control law (3.19) proposed in this chapter), controller 2 (control law (3.41) presented in this chapter), controller 3 [40], and controller 4 [39], in the presence of random Gaussian noise at different levels of $\sigma = 0.01, 0.1, 1$. (a) Controller 1, noise level $\sigma = 1$ and (b) controller 2, noise level $\sigma = 0.01$.

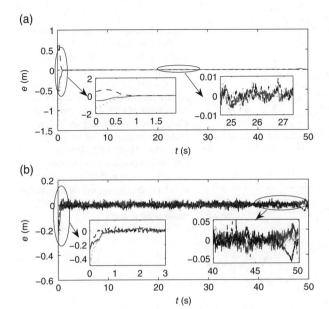

Figure 3.9 Simulation comparisons for the tracking control of the end-effector of a PUMA 560 with respect to a time-varying reference along a circular path using different neuro-controllers, i.e. controller 1 (control law (3.19) proposed in this chapter), controller 2 (control law (3.41) presented in this chapter), controller 3 [40], and controller 4 [39], in the presence of random Gaussian noise at different levels of $\sigma = 0.01, 0.1, 1$. (a) Controller 2, noise level $\sigma = 0.1$ and (b) controller 2, noise level $\sigma = 1$.

of Ω given in (3.52). We simulated the use of (3.19) for regulation to a fixed position with the same parameter configuration as in Section 3.5.1 to verify Theorem 3.1 and the use of (3.41) for the dynamic tracking of a time-varying trajectory with the same parameter configuration as in Section 3.5.2 to verify Theorem 3.2. As shown in Figures 3.5a and 3.6a, the control error converges over time in both cases. In Figures 3.5b and 3.6b, the control actions are either equal to ± 1 or within the small range $[-0.1, 0.1]$, which demonstrates the compliance of the control action with the nonconvex set Ω in (3.52). In both Figures 3.5b and 3.6b, because of the relatively large initial control error, the control actions are as large as ± 1. As time elapses, the control actions reduce to the range of $[-0.1, 0.1]$ after 2 s for Figures 3.5b and after 4 s for Figure 3.6b, and they subsequently remain in this range. The convergence of the control error in Figure 3.6 confirms the effectiveness of Theorems 3.1 and 3.2 for nonconvex projection sets.

3.5.3 Comparisons

In this subsection, we compare the performance of the presented control laws with that of existing RNN solutions for the tracking control of redundant manipulators with time-varying references.

We compare the presented controllers, i.e. controller (3.19) and controller (3.41), with existing controllers based on dynamic neural networks [38–40, 45] for the control of redundant manipulators. The controllers presented in [38, 45] extend the results of [39, 40] regarding velocity-level redundancy resolution to acceleration-level resolution. For

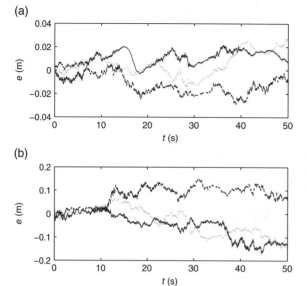

Figure 3.10 Simulation comparisons for the tracking control of the end-effector of a PUMA 560 with respect to a time-varying reference along a circular path using different neuro-controllers, i.e. controller 1 (control law (3.19) proposed in this chapter), controller 2 (control law (3.41) presented in this chapter), controller 3 [40], and controller 4 [39], in the presence of random Gaussian noise at different levels of $\sigma = 0.01, 0.1, 1$. (a) Controller 3, noise level $\sigma = 0.01$ and (b) controller 3, noise level $\sigma = 0.1$.

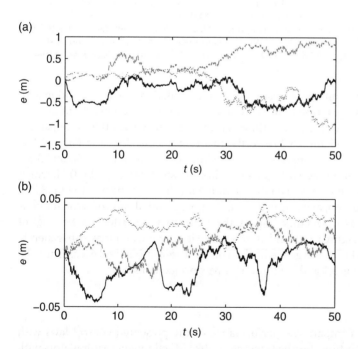

Figure 3.11 Simulation comparisons for the tracking control of the end-effector of a PUMA 560 with respect to a time-varying reference along a circular path using different neuro-controllers, i.e. controller 1 (control law (3.19) proposed in this chapter), controller 2 (control law (3.41) presented in this chapter), controller 3 [40], and controller 4 [39], in the presence of random Gaussian noise at different levels of $\sigma = 0.01, 0.1, 1$. (a) Controller 3, noise level $\sigma = 1$ and (b) controller 4, noise level $\sigma = 0.01$.

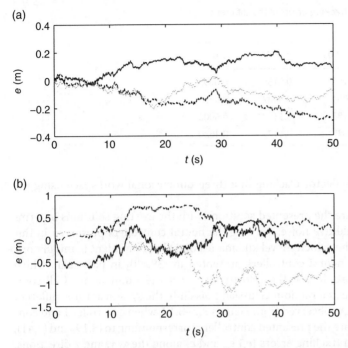

Figure 3.12 Simulation comparisons for the tracking control of the end-effector of a PUMA 560 with respect to a time-varying reference along a circular path using different neuro-controllers, i.e. controller 1 (control law (3.19) presented in this chapter), controller 2 (control law (3.41) presented in this chapter), controller 3 [40], and controller 4 [39], in the presence of random Gaussian noise at different levels of $\sigma = 0.01, 0.1, 1$. (a) Controller 4, noise level $\sigma = 0.1$ and (b) controller 4, noise level $\sigma = 1$.

the controllers presented in this chapter, the focus is on velocity-level resolution. As summarized in Table 3.2, the presented controllers are able to handle a nonconvex projection set Ω, whereas the existing controllers [38–40, 45] require the projection set to be convex. Additionally, because of the lack of direct position feedback in the methods of [38–40 45], they cannot achieve position regulation control with respect to a fixed position. For a time-varying reference position, they require the initial position of the end-effector on the desired reference path for position tracking. By contrast, the two presented controllers are able to cope with both time-invariant regulation and time-varying tracking problems, and they allow the manipulator's end-effector to be initialized at any position in the workspace. Additionally, in the case of additive noise, the existing neural network solutions [38–40, 45] suffer from error accumulation, whereas the presented neural controllers, benefiting from the availability of position feedback, have a bounded error that does not accumulate with time. Another major difference, as indicated in Table 3.2, is that the presented controllers satisfy the set constraint $u(t) \in \Omega, \forall t$. By contrast, the controllers presented in [38–40, 45] are designed to ensure that u falls into the set Ω only after the steady state is reached. Accordingly, a peaking phenomena often occurs in the transition to the steady state. Theoretically, the controllers in [38–40, 45] and those presented in this chapter are all guaranteed to converge. Because of structural differences, they require different numbers of neurons (see Table 3.2) to

Table 3.3 The RMS position tracking errors of the different controllers.

	$\sigma = 0.01$	$\sigma = 0.1$	$\sigma = 1$
Controller 1 (3.19)	0.0102	0.0155	0.0293
Controller 2 (3.41)	0.0034	0.0082	0.0286
Controller 3 [40]	0.0182	0.1123	0.5005
Controller 4 [39]	0.0347	0.1977	0.4630

execute the task of end-effector tracking in a three-dimensional workspace using the PUMA 560.

In addition, we compare the presented solutions with the existing solutions in terms of their robustness to additive noise. Because the neural controllers presented in this chapter are focused on the velocity-level kinematic control of redundant manipulators, we compare them with neural controllers presented in [39, 40] in particular, which also address the same problem at the velocity level. For the simulations of all compared controllers, we injected random Gaussian noise into the system at three different levels of $\sigma = 0.01, 0.1, 1$. As observed from Figures 3.7–3.12, when controller 1 and controller 2 are used, which are the presented controllers corresponding to (3.19) and (3.41), respectively, the position tracking errors (e_x, e_y, and e_z along the x, y, and z directions, respectively) decrease over time and ultimately lie within a bounded range. By contrast, for controller 3 (presented in [40]) and controller 4 (presented in [39]), although the end-effectors are placed on the desired path at time $t = 0$ (as shown in Figures 3.7–3.12, which indicates that all initial errors are zero for controllers 3 and 4), in the presence of additive noise, the tracking errors tend to diverge over time. As the noise level σ increases, the divergence of the tracking errors also increases when controller 3 or 4 is used. As shown in Table 3.3, the root-mean-square (RMS) tracking error between $t = 40$ s and $t = 50$ s is 0.0182 for controller 3 and 0.0347 for controller 4 in the case of $\sigma = 0.01$. These values increase to 0.1123 and 0.1977 for controllers 3 and 4, respectively, when $\sigma = 0.1$ and further increase to 0.5005 for controller 3 and 0.4630 for controller 4 when $\sigma = 1$. By contrast, the RMS tracking errors between $t = 40$ s and $t = 50$ s for controllers 1 and 2 always remain within a bounded range of values that are much lower than the initial tracking errors, thereby confirming their effectiveness.

3.6 Summary

This chapter addresses the control of redundant manipulators using a neural-network-based approach. Two limitations of existing RNN solutions are identified, and modified models are established to overcome these limitations. Rigorous theoretical proofs are supplied to verify the stability of the presented models. Simulation results confirm the effectiveness of the presented solutions and demonstrate their advantages over existing neural solutions.

4

Neural Learning and Control Co-Design for Robot Arm Control

4.1 Introduction

The rapid advances of mechanics, electronics, computer engineering and control theory in recent years have significantly pushed forward the research on manipulators and gained great success in various industrial applications. Redundant manipulators are a special type of manipulators: they have more control degrees of freedom (DOFs) than task DOFs. Redundant manipulators have received intensive research focus for dexterous manipulation of complicated tasks.

A redundant manipulator provides a nonlinear mapping from its joint space to the Cartesian workspace. For the kinematic control of manipulators, it is desirable to find a control action in the joint space such that a reference motion in the workspace can be obtained. The solution to such a problem usually is not unique due to the redundancy and an optimal solution can be reached in terms of certain objective functions and a set of constraints. However, the nonlinearity of manipulators makes it difficult to directly solve the problem in the angle level with a satisfactory accuracy. Instead, most work considers this problem in the velocity space or acceleration space, where the mapping is converted to an affine function, and seek solutions in the new space. Benefiting from the affine nature of the manipulator kinematics in the velocity or acceleration space, early work [25] uses the pseudo-inversion of a manipulator's Jacobian matrix to address the redundancy resolution problem. One by-product of this solution is that it minimizes the kinematic energy consumption. However, this approach introduces intensive overhead for the computation of the pseudo-inverse continuously with time [26]. To avoid the time-consuming computation of the pseudo-inverse, later work [3, 27] removes it by formulating this control problem as a quadratic programming, which minimizes kinematic energy consumption under some physical constraints. Generally, the analytical solution to this formulation cannot be obtained directly. Various matrix manipulations, e.g. matrix decomposition and Gaussian elimination, are employed to seek the numerical solutions [28]. However, the low efficiency of those serial processing techniques prohibits this type of method from real-time control of manipulators.

As all parameters evolve continuously with time in the movement of manipulators, later work designs recurrent neural networks to inherit historically computed solutions. In this way, they can avoid the re-computation of information from scratch and are

Kinematic Control of Redundant Robot Arms Using Neural Networks, First Edition.
Shuai Li, Long Jin and Mohammed Aquil Mirza.
© 2019 John Wiley & Sons Ltd. Published 2019 by John Wiley & Sons Ltd.

able to gain improved performance in real-time computing. In [32], the authors propose a Lagrange neural network model to minimize a cost function under an equation constraint. To deal with inequalities in constrained quadratic programming for manipulator redundancy resolution, researchers turn to consider the problem in its dual space and propose various dual neural network models for accurate solutions. In [34, 39], the authors analyze the problem in the dual space by introducing Lagrange dual variables and derive a convex projection function to represent inequality constraints. Due to the generality of dual neural networks in dealing with constrained quadratic programming problems, they are also extended to solve the redundancy resolution problem based on other alternative models of manipulators, e.g. acceleration space models [3]. To optimize the joint torque, a unified quadratic-programming-based dynamical system approach is established in [38] to control redundant manipulators. A dual neural network with simplified structure is proposed in [40] for the control of kinematically redundant manipulators.

Although great success has been gained for the kinematic control of redundant manipulators using dual neural network approaches, most existing solutions require an accurate knowledge of the manipulator model. This is in contrast to applications in the field of machine learning, where it is usually assumed that the model is unknown when using recurrent neural networks for different tasks. In this chapter, we fill this gap by providing a model-free dual neural network for redundant manipulator control. Different from the pure learning problem of recurrent neural networks, the problem investigated in this chapter involves both learning and control, where they interact with each other.

4.2 Problem Formulation

The forward kinematics of a manipulator is concerned with a nonlinear transformation from a joint space to a Cartesian workspace as described by

$$r(t) = f(\theta(t)), \tag{4.1}$$

where $r(t)$ is an m-dimensional vector in the workspace describing the position and orientation of the end-effector at time t; $\theta(t)$ is an n-dimensional vector in the joint space with each entry describing a joint angle; and $f(\cdot)$ is a nonlinear mapping determined by the structure and the parameters of a manipulator. Due to the nonlinearity and redundancy of the mapping $f(\cdot)$, it is usually difficult to obtain the corresponding $\theta(t)$ directly for a desired $r(t) = r_d(t)$. In contrast, the mapping from the joint space to the workspace in velocity level is an affine one and thus can significantly simplify the problem. To see this, computing the time derivative on both sides of (4.1) yields

$$\dot{r}(t) = J\dot{\theta}(t), \tag{4.2}$$

where $J = \partial f / \partial \theta \in \mathbb{R}^{m \times n}$ is the Jacobian matrix of $f(\cdot)$; and $\dot{r}(t)$ and $\dot{\theta}(t)$ are the Cartesian velocity and the joint velocity, respectively. Each element of $\dot{\theta}(t)$, the angular speed of that particular joint, serves as an input and can be controlled by a motor. Define an n-dimensional input vector $u(t) = \dot{\theta}(t)$. Then, the manipulator kinematics can be rewritten as:

$$\dot{r}(t) = Ju(t). \tag{4.3}$$

Due to physical limitations of motors, the angular speed of joints is limited to a bounded range. For example, the constraint $u(t) \in \Omega = [-\eta, \eta]^n$ applies to the case that each individual joint is restricted to a speed at maximum of η. To capture restrictions on $u(t)$, we set the following constraint for a general situation:

$$u(t) \in \Omega, \tag{4.4}$$

where Ω is a set in an n-dimensional space. For a redundant manipulator described by (4.3) subject to the control input constraint (4.4), we aim to find a control law $u(t)$ such that the velocity tracking error $e(t) = \dot{r}(t) - \dot{r}_d(t)$ for a given reference signal $\dot{r}_d(t)$ converges with time.

4.3 Nominal Neural Controller Design

The kinematic control of redundant manipulators using a dual neural network has been extensively studied in recent years. Although existing methods in this class differ in choosing different objective functions or using different neural dynamics, most of them follow similar design procedures. This class of methods usually formulate the problem as a constrained quadratic optimization, which can be equivalently converted to a set of implicit equations. Then, a convergent dual neural network model with its equilibrium identical to the solution of this implicit equation set is thus devised to solve the problem recursively. Without losing generality, the corresponding optimization problem with the joint space kinematic energy $u^T u = \dot{\theta}^T \dot{\theta}$ as an objective function and with a constraint set to the joint velocity $u \in \Omega$, where superscript u^T denotes the transpose of a vector u (or matrix), can be presented as follows:

$$\min_{u} u^T \Lambda_0 u/2, \tag{4.5a}$$

$$\dot{r}_d = Ju, \tag{4.5b}$$

$$u \in \Omega, \tag{4.5c}$$

where Ω is a convex set and Λ_0 is a positive definite symmetric matrix. In most of the literature [38–40, 45, 46], the set Ω is chosen as $\Omega = \{x \in \mathbb{R}^n, -\eta^- \leq x \leq \eta^+\}$ to capture the physical constraint on the limit of joint speeds. Choosing a Lagrange function as $L(u, \lambda) = u^T \Lambda_0 u/2 + \lambda^T (\dot{r}_d - Ju)$, where $\lambda \in \mathbb{R}^m$ is the Lagrange multiplier corresponding to the equality constraint (4.5b), the optimal solution to (4.5) is equivalent to the solution of the following equation set according to the so-called Karush–Kuhn–Tucker condition [11]:

$$u = P_\Omega \left(u - c_0 \frac{\partial L}{\partial u} \right),$$

$$\dot{r}_d = Ju. \tag{4.6}$$

where $c_0 > 0$ is a constant, $P_\Omega(\cdot)$ is a projection operation to a set Ω which results in a value $y \in \Omega$ such that its distance to x is minimized in the sense of the Euclidean norm, i.e. $P_\Omega(x) = \text{argmin}_{y \in \Omega} \| y - x \|$. A projected recurrent neural network to solve (4.6) can be designed as follows:

$$\varepsilon \dot{u} = -u + P_\Omega \left(u - c_0 \frac{\partial L}{\partial u} \right)$$

$$= -u + P_\Omega(u - c_0 \Lambda_0 u + J^T \lambda), \tag{4.7a}$$

$$\varepsilon \dot{\lambda} = \dot{r}_{\mathrm{d}} - Ju, \tag{4.7b}$$

where $\varepsilon > 0$ is a scaling factor. The dynamics of (4.7) has been rigorously proved to converge to its equilibria that is identical to the optimal solution of (4.5) [39, 40].

4.4 A Novel Dual Neural Network Model

In this section, we present a novel dual neural network model to address the model-free redundancy resolution problem of manipulators.

4.4.1 Neural Network Design

In this section, we design extra neural dynamics to learn the model of the manipulator, and use the online learned results to achieve the model-free control of manipulators.

For the estimation purpose, we first construct a virtual reference $\hat{r}$ with the following dynamic evolution to estimate $\dot{r}$ in the workspace:

$$\dot{\hat{r}} = \hat{J}u \tag{4.8}$$

where $\hat{J}$ is the estimation of the manipulator Jacobian matrix, and $\dot{r} = Ju$ is the measured Cartesian velocity of the end-effector in the workspace. As to $\hat{J}$, we devise the following dynamics for its estimation:

$$\zeta \dot{\hat{J}} = -(\dot{\hat{r}} - \dot{r})u^{\mathrm{T}} = -(\hat{J}u - Ju)u^{\mathrm{T}} \tag{4.9}$$

where $\zeta > 0$ is a scaling factor. By replacing J in (4.7) with $\hat{J}$ obtained in (4.9), the whole system obtained so far can be expressed as follows:

$$\varepsilon \dot{u} = -u + P_{\Omega}(u - c_0 \Lambda_0 u + \hat{J}^{\mathrm{T}} \lambda), \tag{4.10a}$$

$$\varepsilon \dot{\lambda} = \dot{r}_{\mathrm{d}} - \hat{J}u, \tag{4.10b}$$

$$\zeta \dot{\hat{J}} = -(\hat{J}u - \dot{r})u^{\mathrm{T}}. \tag{4.10c}$$

Consider a special case when the initial value of $\hat{J}$ and u at time $t = 0$ are both set at 0. In this situation, the immediate derivative of state variables can be obtained as $\dot{u}(0) = 0$, $\dot{\lambda} = \dot{r}_{\mathrm{d}}(0)$, $\dot{\hat{J}} = 0$, and $\dot{r} = 0$ according to (4.10) and (4.3), implying that $u(t) = 0$, $\hat{J}(t) = 0$, and $r(t) = r(0)$ subsequently for $t > 0$. This reveals the failure of (4.10) for redundancy resolution. The reason lies in the fact that input–output pairs of the manipulator system, i.e. the u–r pairs, lack richness in such a scenario. In other words, there is not enough training data. Intuitively, as the proposed scheme involves both learning and control in a unified framework, it is more challenging than pure learning or identification of manipulator models. However, it also fails even for pure learning of manipulator models when there is only one pair of input–output data available. In this case, only the data pair, namely, $u = 0$ and $r = r(0)$, is available. To avoid this dilemma, we intentionally add extra noises in u to excite the system for the generation of diverging outputs. Specifically, we replace u in (4.10) with $\bar{u}$ for the consistency of notation, define a noise polluted version of u as $\bar{u}$, and propose the following neural adaptive law:

$$\varepsilon \dot{\bar{u}} = -\bar{u} + P_{\Omega}(\bar{u} - c_0 \Lambda_0 \bar{u} + \hat{J}^{\mathrm{T}} \lambda), \tag{4.11a}$$

$$\varepsilon \dot{\lambda} = \dot{r}_{\mathrm{d}} - \hat{J}\overline{u}, \tag{4.11b}$$

$$\zeta \dot{\hat{J}} = -(\hat{J}u - \dot{r})u^{\mathrm{T}}, \tag{4.11c}$$

$$u = \overline{u} + w, \; w \text{ is i.i.d and } \| w \| \leq w_0. \tag{4.11d}$$

where w is a bounded zero mean $\Sigma = \sigma I$ deviation i.i.d. random noise with the bound $w_0 > 0$. About the extra noise w, we have the following remark.

Remark 4.1 In (4.11), the noise w is introduced to increase the diversity of signals. Inevitably this extra noise term also introduces convergence error for redundancy resolution. However, decreasing the bound w_0 helps reduce the impact of w to the convergence error without the reduction of signal diversities. Accordingly, in practice, we may choose a minimal w_0 in terms of the system sensitivity for both effective learning and convergence.

The proposed neural law (4.11) provides a dynamic feedback for the control of a manipulator. In this dynamic feedback mechanism, u, λ, and $\hat{J}$ construct the state variables. u is the output of this neural controller and also serves as the input of the manipulator dynamics (4.3). The expression of the neural law (4.11) does not include any manipulator parameter, which implies that the neural model presented in this chapter is a fully model-free controller. The control is reached based on real-time learning of manipulator characteristics driven by input–output data, i.e. u and r.

Figure 4.1 shows the internal information flow of the proposed scheme. It can be observed that this neural network accepts three input variables, namely the reference speed signal $\dot{r}_d$, the real speed signal $\dot{r}$, and the additive noise w, and outputs u as the control action, which will be further used as the control input of the manipulator. Inside this neural network, the evolution of each of the state variables among u, $\overline{u}$, $\hat{J}$, and λ, depends not only on itself but also on others, which work together to construct the recurrence of the model. Figure 4.2 shows the interconnection of the proposed neural network with a manipulator in a feedback loop for control. Note that the output of the neural network becomes the input of the manipulator and the real-time measurements of the manipulator are then fed back to the neural network as its input, thereby forming a closed loop control.

Figure 4.1 The neural connections between different neural states in the proposed neural controller.

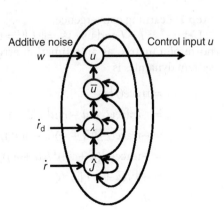

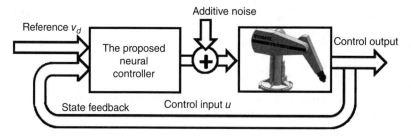

Figure 4.2 The control block diagram of the overall system using the proposed neural network for the control of a manipulator.

With respect to the difference of the proposed model from existing neural controllers for manipulator control, we have the following remark.

Remark 4.2 The proposed neural network is an extension of the existing dual neural network used for manipulator control from model-based to model-free situations. In terms of performance, the proposed model inherits the features of high-accuracy, guaranteed convergence, timely computation, from the dual neural-network-based manipulator control and also creates its own uniqueness, for instance, no need for model information and tolerance to uncertainty.

4.4.2 Stability

In this section, we present the stability theorem of the proposed solution (4.11) and its theoretical proof.

Theorem 4.1 For a redundant manipulator modelled by (4.3) using the proposed neural law (4.11), the model estimation error $\hat{J} - J$ converges to zero and the resolved joint velocity u converges to the optimal redundancy resolution described by (4.5) with an error bounded by w_0 (w_0 can be set arbitrarily small).

Proof: The proof of this theorem is partitioned into three steps: learning convergence; control convergence; and optimality analysis.

Step 1: Learning convergence.
Define $\tilde{J} = \hat{J} - J$ and use $V_1 =\parallel \tilde{J} \parallel_F^2 = \text{trace}(\tilde{J}^{\mathrm{T}}\tilde{J})/2$, where $\parallel \cdot \parallel_F$ denotes the Frobenius norm of a matrix, as an estimation error metric. The time derivative of V_1 along the system dynamics is

$$
\begin{aligned}
\dot{V}_1 &= \text{trace}(\tilde{J}^{\mathrm{T}}\dot{\tilde{J}}) \\
&= -\text{trace}(\tilde{J}^{\mathrm{T}}(\hat{J}u - \dot{r})u^{\mathrm{T}})/\zeta \\
&= -\text{trace}(\tilde{J}^{\mathrm{T}}\tilde{J}(\bar{u} + w)(\bar{u} + w)^{\mathrm{T}})/\zeta \\
&= -\text{trace}((\tilde{J}(\bar{u} + w))^{\mathrm{T}}\tilde{J}(\bar{u} + w))/\zeta \\
&= -\parallel \tilde{J}(\bar{u} + w) \parallel_F^2 /\zeta \\
&\leq 0.
\end{aligned}
\tag{4.12}
$$

Note that $V_1 \geq 0$ and is monotonically decreasing according to (4.12). Therefore, we can apply LaSalle's invariant set principle to find the largest invariant set. Letting $\dot{V}_1 = 0$ yields

$$\tilde{J}(\bar{u} + w) = 0, \quad \text{when } t \to \infty. \tag{4.13}$$

Multiplying $(\bar{u} + w)^\mathrm{T} \tilde{J}^\mathrm{T}$ on both sides of (4.13) and computing the expected value yields

$$
\begin{aligned}
E((\bar{u} + w)^\mathrm{T} \tilde{J}^\mathrm{T} \tilde{J}(\bar{u} + w)) &= E(\text{trace}(\tilde{J}^\mathrm{T} \tilde{J}(\bar{u} + w)(\bar{u} + w)^\mathrm{T})) \\
&= \text{trace}(E(\tilde{J}^\mathrm{T} \tilde{J}\overline{uu}^\mathrm{T})) + \text{trace}(E(\tilde{J}^\mathrm{T} \tilde{J}ww^\mathrm{T})) \\
&\quad + \text{trace}(E(\tilde{J}^\mathrm{T} \tilde{J}\overline{u}w^\mathrm{T})) + \text{trace}(E(\tilde{J}^\mathrm{T} \tilde{J}w\overline{u}^\mathrm{T})) \\
&= \text{trace}(E(\tilde{J}^\mathrm{T} \tilde{J}\overline{uu}^\mathrm{T})) + \text{trace}(E(\tilde{J}^\mathrm{T} \tilde{J})E(ww^\mathrm{T})) \\
&\quad + \text{trace}(E(\tilde{J}^\mathrm{T} \tilde{J}\bar{u})E^\mathrm{T}(w)) + \text{trace}(E(\bar{u}^\mathrm{T} \tilde{J}^\mathrm{T} \tilde{J})E(w)) \\
&= \text{trace}(E(\bar{u}^\mathrm{T} \tilde{J}^\mathrm{T} \tilde{J}\bar{u})) + \sigma^2 \text{trace}(E(\tilde{J}^\mathrm{T} \tilde{J})) \\
&= 0, \quad \text{when } t \to \infty. \tag{4.14}
\end{aligned}
$$

Note that the equality $\text{trace}(E(AB)) = \text{trace}(E(BA))$ for any A and B of appropriate sizes is utilized in the above derivation. Recall that w is an i.i.d. zero mean $\Sigma = \sigma I$ noise. Thus, $\text{trace}(E(\tilde{J}^\mathrm{T} \tilde{J}w\overline{u}^\mathrm{T})) = E(\text{trace}(\tilde{J}^\mathrm{T} \tilde{J}w\overline{u}^\mathrm{T})) = E(\text{trace}(\overline{u}^\mathrm{T} \tilde{J}^\mathrm{T} \tilde{J}w)) = \text{trace}(E(\overline{u}^\mathrm{T} \tilde{J}^\mathrm{T} \tilde{J}w)) = \text{trace}(E(\overline{u}^\mathrm{T} \tilde{J}^\mathrm{T} \tilde{J})E(w)) = 0$. In addition, we have the conclusion that $\text{trace}(E(\tilde{J}^\mathrm{T} \tilde{J}\bar{u}w^\mathrm{T})) = \text{trace}(E(\tilde{J}^\mathrm{T} \tilde{J}\bar{u})E^\mathrm{T}(w)) = 0$. These two equations are both utilized in the derivation of (4.14).

Furthermore notice that $\text{trace}(E(\bar{u}^\mathrm{T} \tilde{J}^\mathrm{T} \tilde{J}\bar{u})) \geq 0$ and $\sigma^2 \text{trace}(E(\tilde{J}^\mathrm{T} \tilde{J})) \geq 0$. Therefore Equation (4.14) implies

$$\text{trace}(E(\tilde{J}^\mathrm{T} \tilde{J})) = E(\text{trace}(\tilde{J}^\mathrm{T} \tilde{J})) = E(\| \tilde{J} \|_F^2) = 0, \quad \text{when } t \to \infty. \tag{4.15}$$

which further implies

$$\tilde{J} = \hat{J} - J = 0, \quad \text{when } t \to \infty. \tag{4.16}$$

Step 2: Control convergence.

The neural dynamics of $\bar{u}$ and λ can be rewritten as follows by replacing $\hat{J}$ with $J + \tilde{J}$:

$$
\begin{aligned}
\epsilon \dot{\bar{u}} &= -\bar{u} + P_\Omega(\bar{u} - c_0 \Lambda_0 \bar{u} + (J + \tilde{J})^\mathrm{T} \lambda), \\
\epsilon \dot{\lambda} &= \dot{r}_\mathrm{d} - (J + \tilde{J})\bar{u}. \tag{4.17}
\end{aligned}
$$

Recall that in step 1 we have proved the convergence of $\tilde{J}$ to zero using the Lyapunov function $V_1 = \text{trace}(\tilde{J}^\mathrm{T} \tilde{J})/2$. In this step, we still employ LaSalle's invariant set principle to conduct convergence analysis in the largest invariant set, i.e. under the condition $\tilde{J} = 0$. In such an invariant set, the neural dynamics reduces to

$$
\begin{aligned}
\epsilon \dot{\bar{u}} &= -\bar{u} + P_\Omega(\bar{u} - c_0 \Lambda_0 \bar{u} + J^\mathrm{T} \lambda), \\
\epsilon \dot{\lambda} &= \dot{r}_\mathrm{d} - J\bar{u}. \tag{4.18}
\end{aligned}
$$

Define a new variable $y = [\bar{u}^\mathrm{T}, \lambda^\mathrm{T}]^\mathrm{T}$. Then, the dynamics of $\bar{u}$ and λ can be represented in terms of y as:

$$\epsilon \dot{y} = -y + P_{\overline{\Omega}}(y - F(y)), \tag{4.19}$$

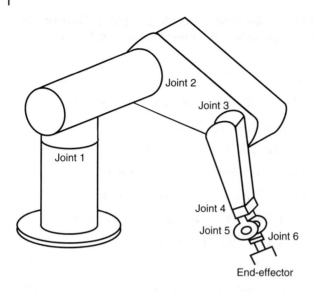

Figure 4.3 A schematic of the PUMA 560 manipulator.

where $\overline{\Omega} = \{(\bar{u}, \lambda), \bar{u} \in \Omega \subset \mathbb{R}^n, \lambda \in \mathbb{R}^m\}$ and

$$F(y) = F(\bar{u}, \lambda) = \begin{bmatrix} c_0 \Lambda_0 \bar{u} - J^T \lambda \\ -\dot{r}_d + J\bar{u} \end{bmatrix} \tag{4.20}$$

with the property that

$$\nabla F = \begin{bmatrix} c_0 \Lambda_0 & -J^T \\ J & 0 \end{bmatrix} \tag{4.21}$$

and

$$\nabla F + \nabla^T F = \begin{bmatrix} 2c_0 \Lambda_0 & 0 \\ 0 & 0 \end{bmatrix} \tag{4.22}$$

which is semi-positive definite. Additionally, for any x and y we have $F(x) - F(y) = \nabla F(z)(x - y)$ with $z = \rho x + (1 - \rho)y$ for certain $0 \le \rho \le 1$ according to the mean-value theorem. Therefore, the following relation always holds:

$$(x - y)^T (F(x) - F(y)) = (x - y)^T \nabla F(z)(x - y) \ge 0 \tag{4.23}$$

which by definition means the mapping $F(\cdot)$ is monotone. According to Theorem 1 in [57], the projected dynamics in (4.19) is Lyapunov stable and globally converges to $y^* = (\bar{u}^*, \lambda^*)$ satisfying

$$(y - y^*)^T F(y^*) \ge 0, \quad \forall y \in \overline{\Omega}. \tag{4.24}$$

That is, for any $\bar{u} \in \Omega$ and $\lambda \in \mathbb{R}^m$,

$$(\bar{u} - \bar{u}^*)^T (c_0 \Lambda_0 \bar{u}^* - J^T \lambda^*) + (\lambda - \lambda^*)^T (-\dot{r}_d + J\bar{u}^*) \ge 0. \tag{4.25}$$

Notice that y in (4.19) converges to the set Ω, i.e. y^* in (4.24) belongs to Ω, as exhibited using a Lyapunov function $V_2 = (y - P_{\overline{\Omega}}(y))^T (y - P_{\overline{\Omega}}(y))/2$, since $\dot{V}_2 = (y - P_{\overline{\Omega}}(y))^T \dot{y} = -(y - P_{\overline{\Omega}}(y))^T (y - P_{\overline{\Omega}}(y - F(y)))/\epsilon \le 0$ and the equality holds when $y \in \overline{\Omega}$.

Table 4.1 Summary of the Denavit–Hartenberg parameters of the PUMA 560 manipulator used in the simulation.

Link	a (m)	α (rad)	d (m)
1	0	$\pi/2$	0.67
2	0.4318	0	0
3	0.4318	$-\pi/2$	0.15005
4	0	$\pi/2$	0
5	0	$-\pi/2$	0
6	0	0	0.2

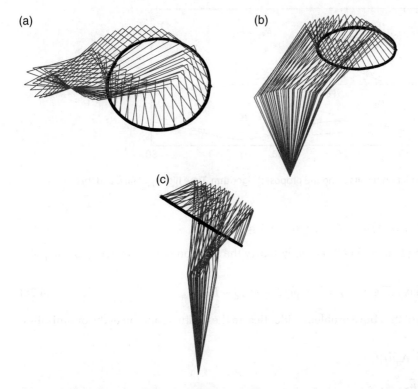

(a)

(b)

(c)

Figure 4.4 The trajectory of the manipulator end-effector using the proposed algorithm with excitation noises, where the piecewise straight lines represent the links of the manipulator and the curve represents the trajectory of the end-effector. (a) x–y view; (b) x–z view; and (c) y–z view.

Step 3: Optimality analysis.

In (4.25), $\lambda \in \mathbb{R}^m$ is allowed to be any value. As a result, we can always choose one such that $(\lambda - \lambda^*)^{\mathrm{T}}(-\dot{r}_d + J\bar{u}^*)$ approaches negative infinity if $-\dot{r}_d + J\bar{u}^* \neq 0$. Accordingly, we conclude that

$$-\dot{r}_d + J\bar{u}^* = 0 \tag{4.26}$$

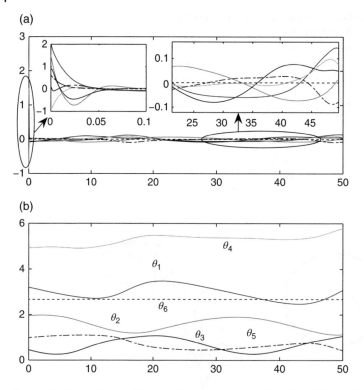

Figure 4.5 Simulation results using the proposed algorithm. Time history of (a) $\bar{u}$ and (b) θ.

and

$$(\bar{u} - \bar{u}^*)^{\mathrm{T}}(c_0 \Lambda_0 \bar{u}^* - J^{\mathrm{T}} \lambda^*) \geq 0, \quad \forall \bar{u} \in \Omega. \tag{4.27}$$

$\bar{u}^*$ and λ^* in (4.26) and (4.27) exactly satisfy the solution of the following saddle point problem

$$\min_{\bar{u} \in \Omega} \max_{\lambda} L(\bar{u} \in \Omega, \lambda) = \bar{u}^{\mathrm{T}} \Lambda_0 \bar{u}/2 + \lambda^{\mathrm{T}}(r_{\mathrm{d}} - J\bar{u}). \tag{4.28}$$

The solution of the above problem is identical to the optimal solution of the optimization problem:

$$\min_{\bar{u}} \bar{u}^{\mathrm{T}} \Lambda_0 \bar{u}/2,$$

$$\dot{r}_{\mathrm{d}} = J\bar{u},$$

$$\bar{u} \in \Omega.$$

Consequently, we conclude that $\bar{u}^*$ and λ^* are the optimal solution of the above optimization problem. Since $u = \bar{u} + w$ is the control input to the manipulator, we thus conclude that u converges to the optimal solution with an error w bounded by w_0. ∎

About this theorem, we have the following remark.

Remark 4.3 In the proposed neural controller (4.11), the noise w is intentionally injected to persistently excite the whole system such that the Jacobian matrix of the

(a)

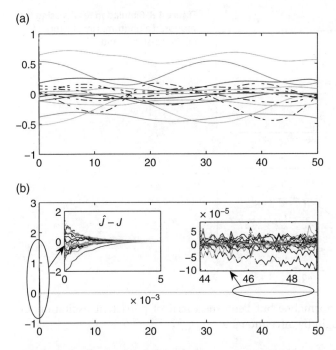

(b)

Figure 4.6 Simulation results using the proposed algorithm. Time history of (a) all elements of the estimated Jacobian matrix $\hat{J}$ and (b) the Jacobian estimation error $\hat{J} - J$.

Figure 4.7 Simulation results using the proposed algorithm. Time history of (a) the position error $r - r_d$ and (b) the resolved velocity error $\dot{r} - \dot{r}_d$.

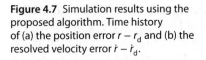

(a)

(b)

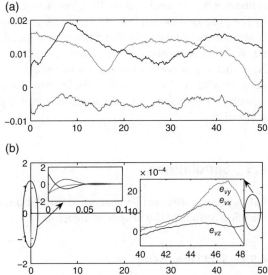

manipulator can be effectively identified at the same time of control. One side-effect of introducing w is the deviation of the solution from the nominal one. However, as proved in Theorem 4.1, the solution error is no more than w_0 in the norm, if $\| w \| \leq w_0$ for any time. This implies that we can always lower the value of $w_0 > 0$ to reduce the

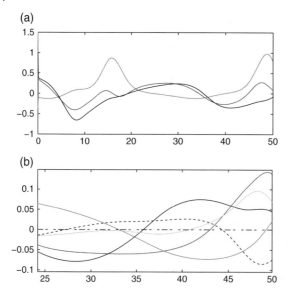

Figure 4.8 Simulation results using the proposed algorithm. Time history of (a) the co-state λ and (b) u.

impact of w to control performance but keep the merit of persistent excitation for Jacobian identification brought by introducing w.

In addition, on the parameter selection in the proposed neural network, we have the following remark.

Remark 4.4 As declared in Theorem 4.1, the convergence of the system controlled by the proposed neural law (4.11) can be guaranteed for $\varepsilon > 0$ and $\zeta > 0$. In fact, the values of ε and ζ matter for the convergence speed. By choosing small ε and ζ, the neural network can be accelerated for convergence, which is thus preferred in implementation. Note that $1/\varepsilon$ serves as the scaling factor for the control part of the proposed neural network while $1/\zeta$ serves that for the learning part. The control part relies on the correct learning for effective control. Therefore, it is also preferred to choose $0 < \zeta \ll \varepsilon \ll 1$ in realization.

4.5 Simulations

In this section, we consider numerical simulations on a PUMA 560 manipulator to show the effectiveness of the proposed algorithms.

4.5.1 Simulation Setup

The PUMA 560 (Figure 4.3) is an articulated manipulator with six independently controlled joints. The parameters of the PUMA 560 manipulator are summarized in Table 4.1. The end-effector of a PUMA 560 manipulator can reach any position at any orientation in its workspace. In this simulation, we particularly consider the position control of the end-effector in three dimensions. In this sense, the 6-DOF

(a)

(b)

(c)

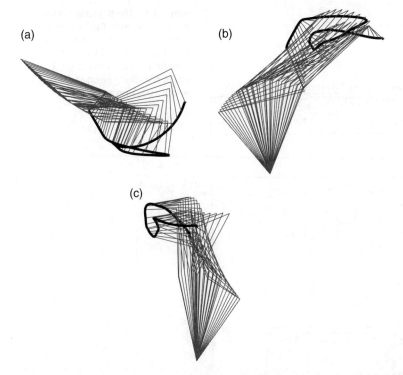

Figure 4.9 The trajectory of the manipulator end-effector using the algorithm without additive noises, where the piecewise straight lines represent the links of the manipulator and the curve represents the trajectory of the end-effector. (a) x–y view; (b) x–z view; and (c) y–z view.

PUMA 560 manipulator is a redundant manipulator relative to this particular task in a three-dimensional space.

In this simulation, the reference end-effector movement is set as a circle. This circle is centered at $(0.35, 0, 1.3)$ with a radius 0.2 m and is inclined around the x-axis for $30°$. The desired angular motion speed is set as 0.2 rad/s. The control scaling factor ε in the neural model is set as $\varepsilon = 10^{-2}$ and the learning scaling factor ζ is set as $\zeta = 10^{-4}$. The excitation signal w is set as a random noise with zero mean, 10^{-3} deviation and the bound $w_0 = 2 \times 10^{-3}$. The goal to set the noise at a small value is to ensure a minimal impact on the system's performance. The projection set Ω is set as $\Omega = [-1, 1]^6$.

4.5.2 Simulation Results

4.5.2.1 Tracking Performance

In this section, we show typical simulations on using the proposed neural network for the tracking of the circular motion. Note that the simulation is conducted under the condition that no prior knowledge about the manipulator parameters is employed. As shown in Figure 4.4, the end-effector of the PUMA 560 manipulator is able to track the desired trajectory successfully based on the proposed neural model with the excitation signal w. The time evolutions of the state variable $\bar{u}$, the joint angle of the manipulator, and the estimated Jacobian matrix (a total of 18 elements) are shown in Figures 4.5a,

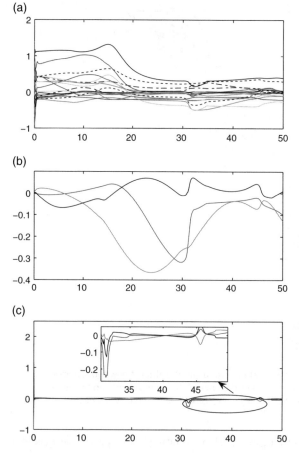

Figure 4.10 The trajectory of the manipulator end-effector using the algorithm without additive noises. Time history of (a) the estimation error for the Jacobian matrix; (b) the position tracking error; and (c) the velocity tracking error.

4.5b, and 4.6a, respectively. As observed, the signals evolve smoothly with time. Figure 4.6b shows the time evolution of $\hat{J} - J$, which captures the learning error. As demonstrated in this figure, the learning error converges to zero very fast and remains a small value at the level of 10^{-5} m/rad after convergence. Recall that the control objective for redundancy resolution is to find a proper value of u such that the resulting end-effector velocity matches the desired reference. With the computed control input, the velocity tracking error also converges to a very small value at the level of 10^{-4} m/s after a short period of transition, as shown in Figure 4.7a. Overall, with the proposed control scheme, both the learning error $\hat{J} - J$ in Figure 4.6b and the position tracking error e in Figure 4.7b converge to zero, which verifies the effectiveness of the proposed algorithm for simultaneous learning and control. The evolutions of the co-state λ and the control action u are shown in Figure 4.8a and b, respectively. Note that u differs from $\bar{u}$ by the additive noises and the curve of $\bar{u}$ in Figure 4.5a almost coincides with that of u since the additive noise w is so small that $\| w \| \leq w_0 = 2 \times 10^{-3} \ll \bar{u}$.

4.5.2.2 With vs. Without Excitation Noises

One key component of the proposed algorithm is the additive noise intentionally injected to the control loop to excite the system for effective learning. As proved in our

Figure 4.11 The position tracking trajectory of the manipulator end-effector under different levels of measurement noises of $\dot{r}$. (a) $\sigma = 0.0005$ and (b) $\sigma = 0.001$.

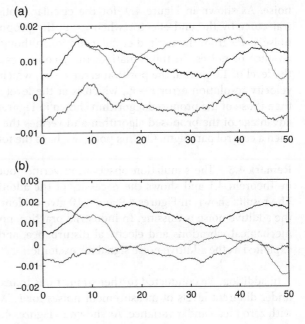

Figure 4.12 The position tracking trajectory of the manipulator end-effector under different levels of measurement noises of $\dot{r}$. (a) $\sigma = 0.005$ and (b) $\sigma = 0.01$.

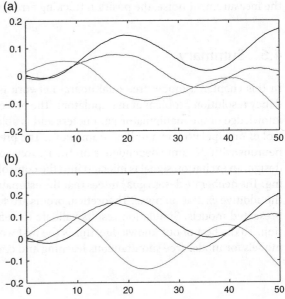

theoretical results and numerically shown in the last section, manipulators with the proposed algorithm applied with excitation noises can successfully reach simultaneous learning and control. In this section, we show in a numerical experiment that the additive noise is necessary to avoid failure of the control task and the performance will degrade greatly after removing the additive noise. To verify this, we re-run the numerical simulation under exactly the same setup as before but removing the additive noise. As shown in Figure 4.9, for the circular motion tracking task, the generated trajectory by the end-effector when using the proposed algorithm without additive noise, is far from the desired circular motion, although it becomes smooth due to the absence of noises. In this situation, the Jacobian estimation error $\hat{J} - J$, which is at the level of 1.0 m/rad, the position error $r - r_d$, which is at the level of 0.1 m, and the velocity resolution error $\dot{r} - \dot{r}_d$, which is at the level of 0.1 m/s, are all much greater than those of the proposed algorithm shown in Figures 4.5–4.8. This demonstrates the advantage of the proposed algorithm and verifies the importance of additive noise in such a control paradigm. On this point, we have the following remark.

Remark 4.5 The simulation observation verifies our theoretical conclusions drawn in Theorem 4.1 and shows the necessity of the additive noise for successful control. The results shown in Figures 4.9 and 4.10 give evidence on the failure of control when the additive noise is missing. In industrial practice, noise occurs in various forms, e.g. mechanical vibrations and electrical disturbance, and is inevitable. Hence, it can be exploited as the additive noise signal for excitation.

In addition, we conducted further simulations to test the performance of the system under different levels of measurement noises for $\dot{r}$. We consider additive white noise with zero mean and σ variance. As shown in Figures 4.11 and 4.12, with the increase of the measurement noise, the position tracking error increases but is always bounded.

4.6 Summary

In this chapter, a model-free dual neural network is presented to address the redundancy resolution problem of manipulators. The presented method does not require any knowledge of the manipulator parameters and is able to address the learning and control of manipulators in a unified framework. The presented neural network consists of neurons with dynamic dependence on their evolutions of state variables. In this novel design, an excitation signal is injected into the control channel for efficient model learning. The deliberate design guarantees that the estimation error converges to zero despite the additive excitation noise. Theoretical proofs are supplied to verify the stability of the presented models. Simulation results validate the effectiveness of the presented solution. To the best of our knowledge, this is the first work to address dual neural network models for model-free simultaneous learning and control of redundant manipulators.

5

Robust Neural Controller Design for Robot Arm Control

5.1 Introduction

In recent years, robotic manipulators have been widely employed in industry for labor-intensive and high-accuracy operations. The current shortage of skilled labor and the aging population pose urgent demands for robotic manipulation, e.g. payload transport, welding, painting, and assembly. Among the broad categories of manipulators, redundant manipulators, which have more control degrees of freedom (DOFs) than desired, are advantageous for dextrous manipulations due to their redundancy in DOFs. In comparison with traditional manipulators without extra DOFs, redundant manipulators usually have more than one control solution for a specific task, and thus allow designers to exploit this feature to fulfil additional requirements, such as obstacle avoidance and control optimization, and thus have received intensive research in recent years.

The introduction of extra DOFs in redundant manipulators increases the capability and flexibility of robotic manipulation, but also sets challenges for the control design for efficient redundancy resolution in real time. Analytically, the joint velocity of a nonredundant manipulator can be expressed in terms of the inverse of its Jacobian matrix. This result extends to redundant manipulators by replacing the Jacobian matrix inverse with its pseudo-inverse, as the Jacobian matrix in this situation is nonsquare [25]. From an optimization perspective, this solution is equivalent to one that minimizes joint speeds. This type of method provides the foundation for manipulator kinematic redundancy but suffers from several drawbacks, e.g. intensive computation due to the continuous computation for pseudo-inverse, outbound of computed joint velocities due to the lack of velocity range constraints, and local instability in some situations due to the singularity of the Jacobian matrix [15]. Later work employs various strategies to deal with the drawbacks of pseudo-inverse-based solutions, including the usage of fast matrix operation algorithms for numerical approximation [28] and iterative methods to approach the solution. However, time efficiency still remains a challenging problem for redundancy resolution since recomputation has to be conducted for every time step.

The parallel processing capability of neural networks inspires researchers to explore the use of neural networks, including feed-forward neural networks [29, 31] and recurrent neural networks [31], to control redundant manipulators. Among them, recurrent

Kinematic Control of Redundant Robot Arms Using Neural Networks, First Edition.
Shuai Li, Long Jin and Mohammed Aquil Mirza.

neural network based solutions achieve great success by exploiting historical information for control update to save computation. In [32], the redundancy resolution problem is modeled as convex optimization under an equation constraint, and solved using a Lagrange neural network. However, this model can only deal with equation constraints. For inequality constraints, e.g. the boundedness of joint velocities, cannot be directly considered by Lagrange neural networks. Dual neural networks are presented to address this problem. In [38], this approach is extended to optimize the joint torque when resolving the redundancy. In [40], the authors propose a single layered dual neural network for the control of kinematically redundant manipulators for the reduction of network spatial complexity and the increase of computational efficiency. There exists an equivalence for redundancy resolutions between the velocity level and the acceleration level. By formulating the problem as constrained quadratic programming at the acceleration level, various dynamic neural solutions can be obtained to reach similar control performances to velocity level resolution results [16]. Not confined to the redundancy resolution of a single serial manipulator, the recurrent neural network approach is applied to kinematic redundancy resolution of parallel Stewart platforms in [58], the coordination of multi-manipulators [3, 42]. Actually, all recurrent neural networks for solving general constrained optimization problems in principle can be applied to address redundancy resolution of manipulators [16].

Although great success has been achieved and various models have been presented for the redundancy resolution of manipulators using recurrent neural networks, most work mainly focuses on the nominal design without considering the appearance of noises explicitly. Actually, in the implementation of neural controllers for manipulator control, the presence of noises is inevitable in the forms of truncation error, rounding error, model uncertainty, and external disturbance. The robustness against additive noises is an important factor in designing reliable neural controllers. However, it remains unexplored for the design of dual neural networks inherently tolerant to noises. This chapter makes progress by proposing a novel recurrent neural network design that is immune to any time-varying noises in the form of polynomials. Based on the feature that polynomial noises become zero after passing through time derivation operations enough times, a neural dynamics system is constructed to learn the noise and counter its impact. Accordingly, we are able to build an improved dual neural network with guaranteed convergence in noisy environments.

5.2 Problem Formulation

In this section, we present the kinematic model of manipulators and formulate the redundancy resolution problem as a control one.

For a k-DOF redundant manipulator with the joint angle $\theta = [\theta_1, \theta_2, ..., \theta_k] \in \mathbb{R}^k$, its Cartesian coordinate $r \in \mathbb{R}^m$ in the workspace can be described as a nonlinear mapping:

$$r = f(\theta). \tag{5.1}$$

For a redundant manipulator, the joint space dimension m is greater than the workspace dimension k, i.e. $m > k$. Equation (5.1) becomes as follows after computing the time derivative on both sides:

$$v = \frac{\partial f}{\partial \theta} w = Jw. \tag{5.2}$$

where $J = \frac{\partial f}{\partial \theta} \in \mathbb{R}^{m \times k}$ is called the Jacobian matrix of f, $v = \dot{r}$ is the end-effector velocity in the workspace, and $w = \dot{\theta}$ is the manipulator joint angular velocity in the joint space. Equation (5.2) describes the kinematics of the manipulator in the velocity level. In comparison with the position level description (5.1), the velocity level descriptions is easier to deal with since $\dot{r}$ is affine to $\dot{\theta}$ in (5.2). In applications, the joint angle usually is driven by acceleration or equivalently by force generated by motors, described as

$$\dot{w} = u + n. \tag{5.3}$$

where u is the acceleration as the control input and, n is an additive noise in the control channel, e.g. the disturbance from electrical and mechanical vibrations. Due to physical constraints, the angular velocity of a manipulator cannot exceed certain limits. Let Ω denote the feasible angular velocity set, then the physical constraint can be expressed as

$$w = \dot{\theta} \in \Omega. \tag{5.4}$$

With (5.4) and (5.2), we are ready to define the kinematic redundancy resolution problem of a manipulator as in the following.

Problem 5.1 (Kinematic Redundancy Resolution) For a redundant manipulator with kinematics (5.2) and a given desired workspace velocity v_d, find a real-time control law $u = h(w, v_d)$ such that the workspace velocity error $v - v_d$ converges to zero, and the joint angular velocity converges to the convex constraint set Ω as in (5.4).

5.3 Dual Neural Networks for the Nominal System

In this section, we present the design of a dual neural network for the control of a nominal manipulator without the presence of noises. This will pave the way for our further consideration with noises taken into account.

5.3.1 Neural Network Design

Due to the redundancy of the manipulator, there is more than one control design that can solve the redundancy resolution. To exploit the redundancy in an optimal fashion, we first define the problem as an optimization one and then convert it into a control law to solve the problem in a recurrent way.

To exploit the extra design freedom, we find a control action to minimize the following quadratic cost function by following conventions for dual neural network design:

$$\min_{w} \frac{1}{2} w^{\mathrm{T}} W w + b^{\mathrm{T}} w, \tag{5.5}$$

where $W \in \mathbb{R}^{k \times k}$ is a symmetric positive semi-definite matrix $W = W^{\mathrm{T}} \geq 0$, $b \in \mathbb{R}^k$ is a vector. This cost function can be utilized to characterize kinematic energy expenses for angular velocity w. For example, choosing $W = I$ as an identity matrix and $b = 0$, this cost function is the kinematic energy. Generally, this cost function describes biased and weighted kinematic energy. The optimal solution to (5.5) also needs to meet the kinematic model (5.2) of the manipulator such that the workspace velocity v equals the desired velocity v_d ultimately, and also the feasibility constraint (5.4). Overall, we formulate it as the following constrained optimization:

$$\min_{w} \frac{1}{2} w^{\mathrm{T}} W w + b^{\mathrm{T}} w, \tag{5.6a}$$

$$v_d = Jw, \tag{5.6b}$$

$$w \in \Omega. \tag{5.6c}$$

Conventionally, a dual neural network is designed based on a Lagrange function as $L = \frac{1}{2} w^{\mathrm{T}} W w + b^{\mathrm{T}} w + \lambda^{\mathrm{T}} (Jw - v_d)$. In this chapter, we augment such a Lagrange function with the equality constraint, which benefits the convergence of the resulting neural dynamics, and define it as follows:

$$
\begin{aligned}
L(w \in \Omega, \lambda) &= \frac{1}{2} w^{\mathrm{T}} W w + \frac{1}{2} (Jw - v_d)^{\mathrm{T}} W' (Jw - v_d), \\
&\quad + b^{\mathrm{T}} w + \lambda^{\mathrm{T}} (Jw - v_d),
\end{aligned}
\tag{5.7}
$$

where W' is a symmetric positive semi-definite weighting matrix $W' = W'^{\mathrm{T}} \geq 0$. The optimal solution to (5.6) is identical to the saddle point of the following problem:

$$\min_{w \in \Omega} \max_{\lambda} L(w, \lambda). \tag{5.8}$$

According to the Karush–Kuhn–Tucker condition [59], the solution of (5.8) satisfies

$$
0 \in \frac{\partial L}{\partial w} + N_\Omega(w),
$$
$$
0 = \frac{\partial L}{\partial \lambda}, \tag{5.9}
$$

where $N_\Omega(w)$ denotes the normal cone of set Ω at w. According to the definition of normal cone, we have

$$s \in N_\Omega(w) \leftrightarrow s^{\mathrm{T}}(x - w) \leq 0, \ \forall x \in \Omega, \tag{5.10}$$

with which, the first expression in (5.9) can be written as

$$-\left(\frac{\partial L}{\partial w}\right)^{\mathrm{T}} (x - w) \leq 0, \ \forall x \in \Omega. \tag{5.11}$$

With (5.11), we further conclude

$$\left\| w - \frac{\partial L}{\partial w} - x \right\|^2 - \left\| \frac{\partial L}{\partial w} \right\|^2 = \| w - x \|^2 + 2\left(\frac{\partial L}{\partial w}\right)^{\mathrm{T}} (x - w) \geq 0, \forall x \in \Omega. \tag{5.12}$$

This implies

$$w = P_\Omega \left(w - \frac{\partial L}{\partial w} \right), \tag{5.13}$$

where $P_\Omega(\cdot)$ is the projection operator defined as

$$P_\Omega(y) = \mathrm{argmin}_{z \in \Omega} \| z - y \|^2, \tag{5.14}$$

or equivalently, $P_\Omega(y) = \alpha$ is the one satisfying

$$\alpha \in \Omega, \text{ and } \| z - y\|^2 - \| \alpha - y\|^2 \geq 0, \forall z \in \Omega. \tag{5.15}$$

The derivation of (5.13) from (5.12) directly follows by setting new variables $z = x$, $y = w - \frac{\partial L}{\partial w}$ and $\alpha = w$ in (5.12). Overall, (5.9) can be finally expressed as

$$w = P_\Omega \left(w - \frac{\partial L}{\partial w} \right)$$
$$= P_\Omega \left(w - Ww - J^T W'(Jw - v_d) - b - J^T \lambda \right),$$
$$0 = \frac{\partial L}{\partial \lambda} = Jw - v_d. \tag{5.16}$$

This is a nonlinear equation set and usually is impossible to solve analytically. To solve it numerically, we propose the following recurrent neural network in the form of

$$\varepsilon \dot{w} = -w + P_\Omega(w - Ww - J^T W'(Jw - v_d) - b - J^T \lambda),$$
$$\varepsilon \dot{\lambda} = Jw - v_d. \tag{5.17}$$

This amounts to the following control law

$$\varepsilon u = -w + P_\Omega \left(w - Ww - J^T W'(Jw - v_d) - b - J^T \lambda \right),$$
$$\varepsilon \dot{\lambda} = Jw - v_d, \tag{5.18}$$

in the absence of noises, i.e. $n = 0$.

5.3.2 Convergence Analysis

About the presented neural network model, we have the following theoretical conclusion.

Theorem 5.1 Suppose there is no noise in the control, i.e. $n = 0$ in (5.3). Neural control law (5.18) solves the kinematic resolution problem of a redundant manipulator modeled by (5.2) and (5.3), and stabilizes the system to the optimal solution of (5.6).

Proof: For the convenience of proof, we define a new variable

$$\eta = \begin{bmatrix} w \\ \lambda \end{bmatrix}, \tag{5.19}$$

and a function

$$F(\eta) = \begin{bmatrix} Ww + J^T W'(Jw - v_d) + b + J^T \lambda \\ -Jw + v_d \end{bmatrix}, \tag{5.20}$$

and an expanded set of Ω as

$$\overline{\Omega} = \{x = [x_1^T, x_2^T]^T, x_1 \in \Omega \subset \mathbb{R}^k, x_2 \in \mathbb{R}^m\}. \tag{5.21}$$

With the assistance of the new variable η, the function $F(\cdot)$ and the set $\overline{\Omega}$, the neural dynamics can be expressed in a compact way as

$$\varepsilon \dot{\eta} = -\eta + P_{\overline{\Omega}}(\eta - F(\eta)). \tag{5.22}$$

This falls into the canonical form of projected dynamic systems. For the function $F(\cdot)$, its gradient is expressed as

$$\nabla F = \begin{bmatrix} W + J^T W' J & J^T \\ -J & 0 \end{bmatrix},$$

which satisfies

$$\nabla F + \nabla^T F = \begin{bmatrix} 2W + 2J^T W' J & 0 \\ 0 & 0 \end{bmatrix} \geq 0.$$

Due to this property, we conclude $F(\cdot)$ is monotone by the following reasoning

$$(\eta_1 - \eta_2)^T(F(\eta_1) - F(\eta_2)) = (\eta_1 - \eta_2)^T \nabla F(\zeta)(\eta_1 - \eta_2)$$

$$= \frac{1}{2}(\eta_1 - \eta_2)^T(\nabla F(\zeta) + \nabla^T F(\zeta))(\eta_1 - \eta_2) \geq 0, \quad \forall \eta_1, \forall \eta_2. \tag{5.23}$$

The projected dynamics in (5.22) is Lyapunov stable and globally converges to $\eta^* = (w^*, \lambda^*)$ satisfying

$$(\eta - \eta^*)^T F(\eta^*) \geq 0, \quad \forall \eta \in \overline{\Omega}. \tag{5.24}$$

Expression (5.24) is a variational inequality and can be equivalently written in the projection form as

$$P_{\overline{\Omega}}(\eta^* - F(\eta^*)) = \eta^*, \tag{5.25}$$

which is the equilibrium point of (5.22). Recall the definition of η, we have

$$P_\Omega(w^* - Ww^* - J^T W'(Jw^* - v_d) - b - J^T \lambda^*) = w^*,$$

$$Jw^* - v_d = 0. \tag{5.26}$$

This is the case when choosing $w = w^*$ in (5.16). The equivalence among (5.16), (5.9), and (5.8) implies that w^* and λ^* are the optimal solution to

$$\min_{w \in \Omega} \max_\lambda L(w, \lambda). \tag{5.27}$$

for $L(w, \lambda)$ defined in (5.7). Since $L(w, \lambda)$ is the Lagrange function of (5.6), we thus conclude the optimality of w^* to the constrained programming (5.6), which completes the proof. ∎

5.4 Neural Design in the Presence of Noises

In the last section, we have designed a neural network to deal with kinematic redundancy resolution in a nominal situation without noises, i.e. $n = 0$ in (5.3). In real applications, noise is inevitable and it is important to design a noise tolerant controller for effective control. In this section, we present the design of neural dynamics that takes noises into account.

5.4.1 Polynomial Noises

In this section, we consider the situation with polynomial noises. According to Taylor expansion, it is possible to represent any time varying noises by a polynomial for a desired accuracy. In this sense, a polynomial noise can be regarded as a general representation of noises.

5.4.1.1 Neural Dynamics

We first present the neural dynamics with the capability to deal with polynomial noises. For an l-order polynomial noise $n = \sum_{i=0}^{l} n_i$ with $n_i = k_i t^i$ for $i = 0, 1, ..., l$ and k_i unknown, we present the neural controller as follows:

$$\varepsilon u = -(c_0 \varepsilon + 1)w + P_\Omega(w - Ww - J^T W'(Jw - v_d))$$

$$- b - J^T \lambda) - \varepsilon \sum_{i=1}^{l} c_i \eta_i - \sum_{i=0}^{l} c_i \mu_i,$$

$$\varepsilon \dot{\lambda} = Jw - v_d,$$

$$\dot{\mu}_0 = w - P_\Omega(w - Ww - J^T W'(Jw - v_d) - b - J^T \lambda),$$

$$\dot{\mu}_i = \mu_{i-1} \quad \forall i = 1, 2, ..., l,$$

$$\dot{\eta}_1 = w,$$

$$\dot{\eta}_i = \eta_{i-1} \quad \forall i = 2, 3, ..., l. \tag{5.28}$$

About this neural controller, we have the following theoretical results.

Theorem 5.2 Suppose the noise in (5.3) is an l-order polynomial relative to time, i.e. $n = \sum_{i=0}^{l} n_i$ with $n_i = k_i t^i$ for $i = 0, 1, ..., l$, where k_i is an unknown constant and t represents time. Neural control law (5.28) solves the kinematic resolution problem of a redundant manipulator modeled by (5.2) and (5.3), and stabilizes the system to the optimal solution of (5.6), provided the coefficients c_i for $i = 0, 1, 2, ..., l$ are chosen such that all roots of the polynomial $s^{l+1} + c_0 s^l + c_1 s^{l-1} + ... + c_l = 0$ are in the left half plane, no matter how large the constant k_i is for $i = 1, 2, ..., l$.

Proof: We first define the control error as $e = w - P_\Omega(w - Ww - J^T W'(Jw - v_d) - b - J^T \lambda)$. Then, after substituting the expression of u in (5.18) for n, the neural dynamics (5.28) becomes

$$\varepsilon \dot{w} = -e - c_0 \varepsilon w - \varepsilon \sum_{i=1}^{l} c_i \eta_i - \sum_{i=0}^{l} c_i \mu_i + \varepsilon \sum_{i=0}^{l} n_i,$$

$$\varepsilon \dot{\lambda} = Jw - v_d,$$

$$\dot{\mu}_0 = e,$$

$$\dot{\mu}_i = \mu_{i-1} \quad \forall i = 1, 2, ..., l,$$

$$\dot{\eta}_1 = w,$$

$$\dot{\eta}_i = \eta_{i-1} \quad \forall i = 2, 3, ..., l. \tag{5.29}$$

Define a new variable $p_i = \varepsilon \eta_i + \mu_i$ for $i = 1, 2, ..., l$, and $p_0 = \varepsilon w + \int_0^t e \, dt$. Then, $\dot{p}_i = \varepsilon \dot{\eta}_i + \dot{\mu}_i = \varepsilon \eta_{i-1} + \mu_{i-1} = p_{i-1}$ for $i = 1, 2, ..., l$. With the new variable p_i, (5.29) can be further simplified to

$$\dot{p}_0 = -\sum_{i=0}^{l} c_i p_i + \varepsilon \sum_{i=0}^{l} n_i,$$

$$\dot{p}_i = p_{i-1} \quad \forall i = 1, 2, 3, ..., l. \tag{5.30}$$

It is noteworthy that the above dynamics, in terms of the state variables p_i for $i = 0, 1, 2, ..., l$, represents a linear system with the noise $\varepsilon \sum_{i=0}^{l} n_i$ as input. We apply

Laplace transformation to explore the stability and steady-state values of such a linear system with polynomial inputs. After transformation, the frequency domain equivalence to (5.30) is obtained as

$$sP_0(s) - p_0(0) = -\sum_{i=0}^{l} c_i P_i(s) + \varepsilon \sum_{i=0}^{l} N_i(s),$$
$$sP_i(s) - p_i(0) = P_{i-1}(s) \quad \forall \ i = 1, 2, 3, ..., l. \tag{5.31}$$

where $P_i(s) = \mathcal{L}(p_i(t))$ represents the Laplace transform of p_i for $i = 0, 1, 2, ..., l$. $N_i(s) = \mathcal{L}(n_i(t)) = \mathcal{L}(k_i t^i) = i!/s^{i+1}$, where $i! = \Pi^i_{j=1} j$ represents the Laplace transform of $n_i = k_i t^i$. From the second equation in (5.31), it is obtained that

$$P_i(s) = \frac{p_i(0)}{s} + \frac{p_{i-1}(0)}{s^2} + \frac{p_{i-2}(0)}{s^3} + ... + \frac{p_1(0)}{s^i} + \frac{P_0(s)}{s^i}. \tag{5.32}$$

Submitting (5.32) into the first equation of (5.31) results in

$$sP_0(s) - p_0(0) = -P_0(s)\left(c_0 + \frac{c_1}{s} + \frac{c_2}{s^2} + ... + + \frac{c_l}{s^l}\right)$$
$$- \frac{1}{s}(c_1 p_1(0) + c_2 p_2(0) + ... + c_i p_i(0))$$
$$- \frac{1}{s^2}(c_2 p_1(0) + c_3 p_2(0) + ... + c_i p_{i-1}(0))$$
$$...$$
$$- \frac{1}{s^l} c_l p_1(0) + \varepsilon \sum_{i=0}^{l} N_i(s). \tag{5.33}$$

After reorganization, the above equation becomes

$$P_0(s) = \frac{1}{s^{l+1} + c_0 s^l + c_1 s^{l-1} + ... + c_l}(p_0(0)s^l$$
$$- s^{l-1}(c_1 p_1(0) + c_2 p_2(0) + ... + c_l p_l(0))$$
$$- s^{l-2}(c_2 p_1(0) + c_3 p_2(0) + ... + c_l p_{l-1}(0))$$
$$... - c_l p_1(0))$$
$$+ \frac{\varepsilon s^l \sum_{i=0}^{l} N_i(s)}{s^{l+1} + c_0 s^l + c_1 s^{l-1} + ... + c_l}. \tag{5.34}$$

Now we define $y = p_0$. In the frequency domain $Y(s) = \mathcal{L}(y(t)) = sP_0(s) - p_0(0)$. It can be further expressed as

$$Y(s) = \frac{s}{s^{l+1} + c_0 s^l + c_1 s^{l-1} + \dots + c_l}(p_0(0)s^l$$
$$- s^{l-1}(c_1 p_1(0) + c_2 p_2(0) + \dots + c_l p_l(0))$$
$$- s^{l-2}(c_2 p_1(0) + c_3 p_2(0) + \dots + c_l p_{l-1}(0))$$
$$\dots - c_l p_1(0) - p_0(0))$$

$$+ \frac{\varepsilon s^{l+1} \sum_{i=0}^{l} N_i(s)}{s^{l+1} + c_0 s^l + c_1 s^{l-1} + \dots + c_l}. \tag{5.35}$$

Consider the system with $\sum_{i=0}^{l} N_i(s)$ as input and $Y(s)$ as output. Its transfer function can be defined as $G(s) = Y(s)(\sum_{i=0}^{l} N_i(s))^{-1}$, which characterizes the mapping from the noise $\sum_{i=0}^{l} N_i(s)$ to $Y(s)$ in the frequency domain. From (5.35), it is clear that the transfer function $G(s)$ can be written as

$$G(s) = \frac{\varepsilon s^{l+1}}{s^{l+1} + c_0 s^l + c_1 s^{l-1} + \dots + c_l}. \tag{5.36}$$

The system characterized by $G(s)$ is stable because all roots of its characteristic polynomial $s^{l+1} + c_0 s^l + c_1 s^{l-1} + \dots + c_l = 0$ are in the left half plane according to the Routh–Hurwitz criterion. In Equation (5.34), the term $s(p_0(0)s^l - s^{l-1}(c_1 p_1(0) + c_2 p_2(0) + \dots + c_l p_l(0)) - s^{l-2}(c_2 p_1(0) + c_3 p_2(0) + \dots + c_l p_{l-1}(0))\dots - c_l p_1(0))/(s^{l+1} + \dots + c_l)$ corresponds to the zero input response and the term $\varepsilon s^{l+1} \sum_{i=0}^{l} N_i(s)/(s^{l+1} + c_0 s^l + c_1 s^{l-1} + \dots + c_l)$ to the zero state response. The employment of final value theorem to (5.34) yields

$$\lim_{t \to \infty} y(t) = \lim_{s \to 0} sY(s)$$
$$= \lim_{s \to 0} \frac{s^2}{s^{l+1} + c_0 s^l + c_1 s^{l-1} + \dots + c_l}(p_0(0)s^l$$
$$- s^{l-1}(c_1 p_1(0) + c_2 p_2(0) + \dots + c_l p_l(0))$$
$$- s^{l-2}(c_2 p_1(0) + c_3 p_2(0) + \dots + c_l p_{l-1}(0))$$
$$\dots - c_l p_1(0) - p_0(0)$$
$$+ \varepsilon \sum_{i=0}^{l} \frac{s^i i!}{s^{i+1}} \bigg) = 0. \tag{5.37}$$

Recall that $y = \dot{p}_0 = \varepsilon \dot{w} + d(\int_0^t e \, dt)/dt = \varepsilon \dot{w} + e = \varepsilon \dot{w} + w - P_\Omega(w - Ww - J^T W'(Jw - v_d) - b - J^T \lambda) = 0$, which implies that the system dynamics ultimately reduces to the nominal system controlled by the dual neural network without the perturbation of any noises. According to Theorem 5.1, w solves the kinematic redundancy resolution problem and stabilizes to the optimal solution of (5.6), which completes the proof. ∎

About the choice of parameters c_i for $i = 0, 1, 2, \dots, l$ in the neural controller, we have the following remark.

Remark 5.1 The neural coefficients are required to be chosen such that all roots of the polynomial $s^{l+1} + c_0 s^l + c_1 s^{l-1} + \ldots + c_l = 0$ are in the left half plane. As an easy way for their choice, we may set c_i as follows:

$$c_0 = (l+1)s_0,$$

$$c_1 = \frac{(l+1)l}{2}s_0^2,$$

$$c_2 = \frac{(l+1)l(l-1)}{6}s_0^3,$$

$$\ldots$$

$$c_i = \frac{(l+1)!}{(l-i)!(i+1)!}s_0^{i+1},$$

$$\ldots$$

$$c_l = s_0^{l+1}, \tag{5.38}$$

which are coefficients of the binomial $(s + s_0)^{l+1} = \sum_{i=0}^{l+1} \frac{(l+1)!}{i!(l+1-i)!} s_0^{l+1-i} s^i = 0$ for $s_0 \in \mathbb{R}$ and $s_0 > 0$. Note that the roots of this binomial are all $-s_0$ and the root condition is satisfied automatically.

About the generality of the presented neural controller in coping with any type of time-varying noises, we have the following remark.

Remark 5.2 Taylor's theorem claims that it can be reached for any desired accuracy to approximate a nonlinear function with polynomial Taylor series under some mild conditions. In other words, we can always find an l to represent the noise as an l-order polynomial noise $n = \sum_{i=0}^{l} n_i$ with $n_i = k_i t^i$ with an arbitrarily small approximation error. This implies the generality of the presented neural controller (5.28) to deal with arbitrary noises.

The presented method in this chapter is applicable to various real problems. On this point, we have the following remark.

Remark 5.3 In real applications, especially industrial ones, manipulators are inevitably affected by additive noises and this often results in the deviation of the real trajectory of the end-effector from the desired one. The presented method in this chapter significantly improves the tolerance of neural controllers to noise and thus is applicable to various engineering applications using robotic manipulators, e.g. painting robots, welding robots, quadlegged robots, and humanoid robots. The model considered in this chapter is a general one and it is suitable for any type of serial redundant-manipulators with any number of DOFs. The current chapter mainly focuses on the theoretical part, i.e. theoretical derivation of the neural controllers with high tolerance to additive noise, with its verification using numerical simulations based on PUMA 560 robotic arms. The verification on real robotic arms is beyond the scope of the current chapter and will be covered in our future research that will emphasize the practical implementation on real robotic hardware.

Figure 5.1 The architecture of the presented recurrent neural networks.

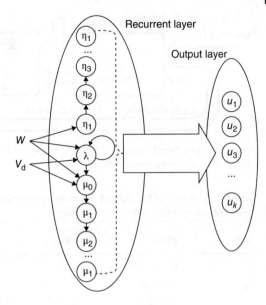

5.4.1.2 Practical Considerations

In real applications, it is necessary to determine a proper order for the polynomial noise before implementing the neural controller. For a particular problem, a trial-and-error procedure can be used to determine its value. Specifically, we can first assume the polynomial order is $l = 0$, i.e. a constant noise, and design neural controllers accordingly. With the devised neural controller, we can measure the control performance in terms of the tracking error. If it is not satisfactory, we can increase the polynomial order to $l = 1$, and check the tracking error again. Otherwise, we just stop iteration and use the current neural controller for control. If $l = 1$ cannot return with a satisfactory performance, we proceed to try $l = 2$. Recursively, we can find a proper order of the polynomial, and the corresponding neural controller, with which the control error is satisfactory. Note that we can always improve the tracking accuracy by increasing the order l as implied by Taylor's theorem on the approximation power of polynomials.

In addition, we present the architecture of the presented neural network (Figure 5.1) to show its structure from a connectionist perspective. It is clear from equation (5.28) and Figure 5.1 that the neural model contains hidden states $\lambda, \mu_0, \mu_1, \mu_2, \ldots, \mu_l, \eta_1, \eta_2, \ldots,$ η_l and an explicit output u. The hidden states interact with each other in a dynamic way and form the recurrent layer of this neural network. The recurrent layer then impacts the output u with a static mapping and forms the output layer. The dependency of variables and the relationship between them as described by (5.28) are shown in Figure 5.1. From this figure, it can also be observed that there are $2l + 2$ variables in the recurrent layer, i.e. $\lambda, \mu_0, \mu_1, \mu_2, \ldots, \mu_l, \eta_1, \eta_2, \ldots, \eta_l$. As to η, w impacts η_1, η_1 impacts η_2, ..., and η_{l-1} impacts η_l, in a cascaded way. The hidden variable λ dynamically depends on itself and the two inputs w and v_d. As to μ, the dynamics of μ_0 depends on λ, w and v_d, the dynamics of μ_1 only depends on μ_0, the dynamics of μ_2 only depends on μ_1, ..., and the dynamics of μ_l only depends on μ_{l-1}. All variables in the recurrent layer then impact the output u in a nonlinear way.

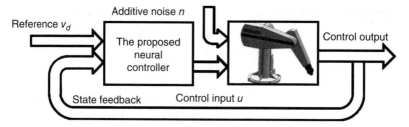

Figure 5.2 The control diagram of the system using the presented neural controller to steer the motion of a redundant manipulator.

Table 5.1 Summary of the Denavit–Hartenberg parameters of the PUMA 560 manipulator used in the simulation.

Link	a(m)	α(rad)	d(m)
1	0	$\pi/2$	0
2	0.43180	0	0
3	0.02030	$-\pi/2$	0.15005
4	0	$\pi/2$	0.43180
5	0	$-\pi/2$	0
6	0	0	0.20000

Furthermore, the implementation of the interaction between the neural controller and the manipulator is also very important in real implementations. The manipulator is driven by the acceleration u of joint motors. With the actuation of all individual joints, the end-effector moves in its workspace and the control objective is to reach the desired velocity v_d in the workspace. For the neural controller in the presence of polynomial noises, the input to this neural controller includes two quantities: one is w, which is the angular velocity of joints and comes from the feedback provided by sensors mounted on the manipulator; and the other one is the reference velocity v_d, which comes from the control command. As a summary, we provide the control diagram in Figure 5.2 to describe the interconnection between the presented neural controller and the manipulator.

5.4.2 Special Cases

In this section, we examine two special cases of polynomial noises, namely constant noises and linearly time-varying noises, to show the specific presentation of the neural controllers.

5.4.2.1 Constant Noises

We first consider the situation with a constant noise. In practice, this category of noises can be employed to describe uncompensated weights, manipulator parameter errors, etc.

(a)

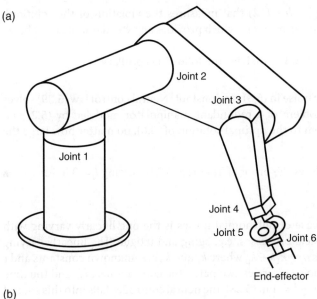

(b)

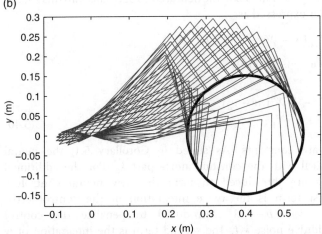

Figure 5.3 The schematic of a 6-DOF PUMA 560 manipulator considered in the simulation and the generated trajectory of PUMA 560 controlled by the presented neural network in the absence of noises. (a) PUMA 560 manipulator and (b) end-effector trajectory.

Choosing $l = 0$ in (5.28), the neural controller falls into the special case with a constant noise, as represented in the following form

$$\varepsilon u = -(c_0\varepsilon + 1)w + P_\Omega\left(w - Ww - J^\mathrm{T}W'(Jw - v_d)\right)$$
$$-b - J^\mathrm{T}\lambda) - c_0\mu,$$
$$\varepsilon\dot\lambda = Jw - v_d,$$
$$\dot\mu = w - P_\Omega\left(w - Ww - J^\mathrm{T}W'(Jw - v_d) - b - J^\mathrm{T}\lambda\right). \qquad (5.39)$$

In comparison with the neural controller (5.18) in the absence of noises, two additional terms are inserted into the control law: the term $-c_0\mu$, where μ is the integration of the

$w - P_\Omega(w - Ww - J^T(Jw - v_d) - b - J^T\lambda)$ that measures the violation of the solution's optimality in quantity, and the term $-c_0\varepsilon w$, which provides an extra damping to stabilize the system.

About the neural controller (5.39), we have the following corollary.

Corollary 5.1 Suppose the noise in (5.3) is a constant. Neural control law (5.39) solves the kinematic resolution problem of a redundant manipulator modeled by (5.2) and (5.3), and stabilizes the system to the optimal solution of (5.6), no matter how large the constant noise is.

Proof: This result directly follows the proof of Theorem 5.2 by setting $l = 0$ in it. ∎

5.4.2.2 Linear Noises

Another important special case of polynomial noises is the one linearly varying with time. It models various physical processes, e.g. aging and fatigue. The linearly varying noise n can be represented as $n = k_1 t + k_0$ where k_1 and k_0 are unknown constants and t represents time. The linear noise n carries two parts: the constant part k_0; and the time proportional part $k_1 t$. Choosing $l = 1$ in (5.28), the neural controller falls into this special case with its neural controller expressed as:

$$\varepsilon u = -(c_0\varepsilon + 1)w + P_\Omega\left(w - Ww - J^T W'(Jw - v_d)\right)$$
$$-b - J^T\lambda) - c_0\mu_0 - c_1\varepsilon\theta - c_1\mu_1,$$
$$\varepsilon\dot\lambda = Jw - v_d,$$
$$\dot\mu_0 = w - P_\Omega\left(w - Ww - J^T W'(Jw - v_d) - b - J^T\lambda\right),$$
$$\dot\mu_1 = \mu_0, \tag{5.40}$$

where $c_0 > 0$ and $c_1 > 0$ are constants. As stated in Corollary 5.1, the neural model ((5.39)) is able to deal with the constant noise part k_0. For the additional part $k_1 t$, two extra terms are introduced to form the new neural controller: $-c_1\mu_1$ and $-c_1\varepsilon\theta$. The first term is the twice integration of the control error $w - P_\Omega\left(w - Ww - J^T W'(Jw - v_d) - b - J^T\lambda\right)$ introduced to penalize the control error resulting from the additive noise $k_1 t$. The second term is the integration of w introduced as a damping term for stabilization consideration. Notice that the newly introduced terms are both in integration format, due to the fact that the time-varying noise $n = k_1 t + k_0$ approaches infinity with the increase of time t and an increased amount of control action is desired to conquer its impact.

About the presented neural controller (5.40), we have the following convergence result in the presence of linear noises.

Corollary 5.2 Suppose the noise in (5.3) is linear in time, i.e. $n = k_1 t + k_0$ where k_1 and k_0 are unknown constants and t represents time. Neural control law (5.40) solves the kinematic resolution problem of a redundant manipulator modeled by (5.2) and (5.3), and stabilizes the system to the optimal solution of (5.6), no matter how large the constants k_1 and k_0 are.

Proof: This result directly follows the proof of Theorem 5.2 by setting $l = 1$ in it. ∎

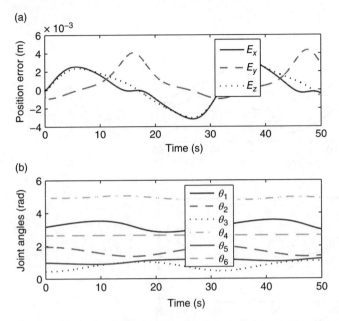

Figure 5.4 The time profile of control parameters for the presented neural network in the absence of noises. (a) Position tracking error and (b) joint angles.

5.5 Simulations

In this section, we apply the presented neural controller to the redundancy resolution of a PUMA 560 robot arm for the tracking of a circular motion. We compare the presented method with existing dynamic neural solutions in the presence of various noises and show its robustness in dealing with additive noises.

5.5.1 Simulation Setup

In the simulation, we consider a PUMA 560 robot arm, with its Denavit–Hartenberg parameters summarized in Table 5.1, as the manipulator platform for the tracking of a circular motion with its end-effector. PUMA 560 is a serial manipulator with six joints and thus has 6 DOFs for control. As the control task only refers to the position tracking, it forms a three-dimensional workspace. For this task, PUMA 560 serves as a redundant manipulator. The desired motion is to track a circle in the $x - y$ plane with radius 0.15 m at an angular speed of 0.2 rad/s. In the simulation, we consider noises up to fourth order of polynomials with different amplitude. For parameters in the presented neural network, we choose $\varepsilon = 10^{-2}$, $W = I$, $W' = 0$, and $b = 0$. As to the parameter c_i, it is determined by Equation (5.38) by choosing $s_0 = 2$. The simulation results are compared with existing neural solutions presented in [39] and [60] by setting the same parameters.

5.5.2 Nominal Situation

In this section, we present simulation results in the nominal situation without the presence of noises to show the convergence of the presented neural controller. Without

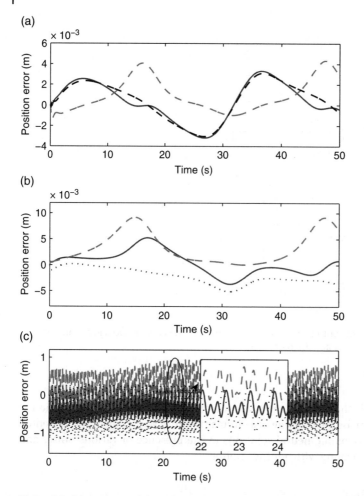

Figure 5.5 Comparisons of the presented approach with existing ones in the presence of different levels of noises with noise $n = 0.1/\varepsilon$. (a) This chapter; (b) [60]; and (c) [39].

losing generality, we consider the situation with $l = 5$. Recall that we have chosen $s_0 = 2$ and we have $c_0 = 12, c_1 = 60, c_2 = 160, c_3 = 240, c_4 = 192$, and $c_5 = 64$ from (5.38). Figures 5.3 and 5.4 show the result of a typical simulation run for 50 s. As shown in Figure 5.3, the generated trajectory for the PUMA 560 manipulator matches the desired circular motion. Quantitatively, the position tracking error in the workspace along x, and y and z directions are drawn in Figure 5.4a. As observed from this figure, the position tracking error remains at a low level (lower than 5×10^{-3} in absolute values for all dimensions). The generated joint angles at different times are shown in Figure 5.4b.

5.5.3 Constant Noises

In this simulation, we set the noise as a constant, and compare the performance obtained by using different neural controllers. Note that the noise goes into the control channel in the form of (5.3) and the control input u is amplified by $1/\varepsilon$. Therefore, we consider

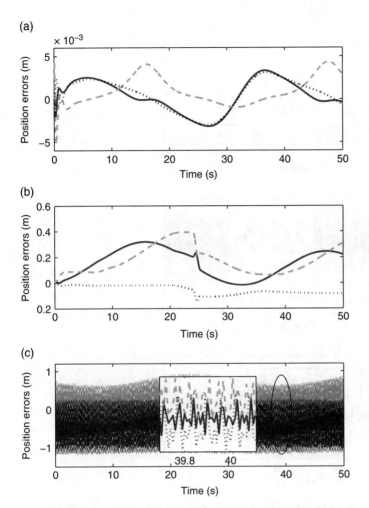

Figure 5.6 Comparisons of the presented approach with existing ones in the presence of different levels of noises with noise $n = 1/\varepsilon$. (a) This chapter; (b) [60]; and (c) [39].

a set of noise comparable with the control input as $n = 0.1/\varepsilon$, $n = 1/\varepsilon$, and $n = 10/\varepsilon$. We compare the results generated by the presented algorithm in this chapter with two existing neural controllers presented in [60] and [39] with the same parameter setup. As shown in Figures 5.5–5.7, with the presence of the constant noise $n = 0.1/\varepsilon$, the position tracking error for the presented method is restricted to be lower than 5×10^{-3} in the absolute value across the entire simulation run. In contrast, the dual neural method presented in [60] has a larger tracking error with a peak value around 10×10^{-3} at time $t = 15$ s and $t = 48$ s (Figure 5.6). As to the dual neural method presented in [39], the system loses stability and fluctuates severely with time. With the increase of the noise to $n = 1/\varepsilon$, the position tracking error of the method presented in [60] increases a lot and reaches a peak value 0.4 at time $t = 25$ s, and the system controlled by the neural law presented in [39] demonstrates enhanced fluctuation with an increased frequency. Note that due to the restriction introduced by the physical size of the manipulator, the

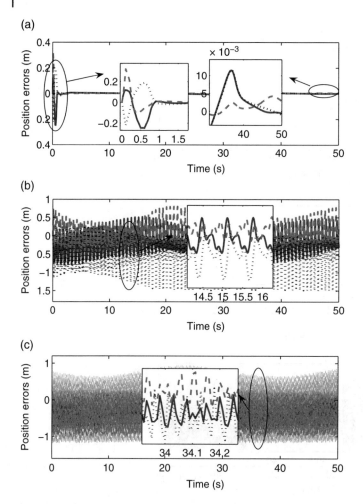

Figure 5.7 Comparisons of the presented approach with existing ones in the presence of different levels of noises with noise $n = 10/\varepsilon$. (a) This chapter; (b) [60]; and (c) [39].

position tracking error is always bounded. For the case with noise $n = 1/\varepsilon$, in contrast to the severe performance degradation of existing methods [39, 60], the presented one still performs well with a similar noise level as in the situation without the presence of noise. For even larger noise $n = 10/\varepsilon$, the method presented in [60] also loses stability and starts to oscillate with a big error amplitude as observed in Figure 5.7. As to the presented method shown in Figure 5.7, after a short period for transition, the position tracking error returns to a small value below 10×10^{-3} very soon. Note that the short transition shown in the left inset figure in Figure 5.7a is due to the fact that the large constant noise $n = 10/\varepsilon$ acts on the system at the beginning of this simulation run while it takes a short transient time for the presented control to adapt to this noise. Overall, the simulation reveals that existing solutions designed in nominal situations are not robust against additive noises and the presented neural controller with an explicit consideration

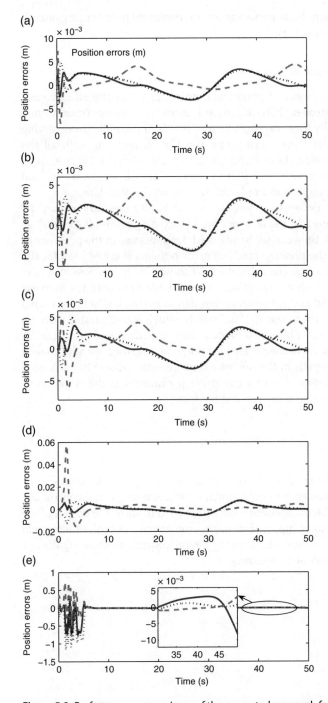

Figure 5.8 Performance comparisons of the presented approach for the tracking of circular motions in the presence of various noises. (a) Linear noise $n = 10t/\varepsilon$; (b)quadratic noise $n = 10t^2/\varepsilon$; (c) cubic noise $n = 10t^3/\varepsilon$; (d) fourth-order noise $n = 10t^4/\varepsilon$; and (e) noise $n = 10t^4/\varepsilon + 10t^3/\varepsilon + 10t^2/\varepsilon + 10t/\varepsilon + 10/\varepsilon$.

of noise tolerance receives a significant performance improvement over existing ones in dealing with unknown constant noises.

5.5.4 Time-Varying Polynomial Noises

In this simulation, we consider a set of polynomial noises to verify the effectiveness of the presented neural controller. Particularly, we consider a linear time-varying noise $n = 10t/\varepsilon$, a quadratically time-varying one $n = 10t^2/\varepsilon$, a cubically time-varying one $n = 10t^3/\varepsilon$, a fourth-order time-varying one $n = 10t^4/\varepsilon$, and one with all the above ingredients plus a constant, i.e. $n = 10t^4/\varepsilon + 10t^3/\varepsilon + 10t^2/\varepsilon + 10t/\varepsilon + 10/\varepsilon$. Notice that the noise signals considered in this section are all time-varying and their values could be very large. For example, at the end of the simulation $t = 50$ s, the linear time-varying noise becomes $n = 500/\varepsilon$, which is 50 times larger than the largest constant noise considered in the last section. For existing methods [39, 60], as demonstrated in Section 5.5.3, they cannot handle such large noises. In the presence of the above mentioned noises, the position tracking error remains at a low level for the linear noise, the quadratic noise and the cubic noise as shown in Figures 5.8a, b, and c, respectively. Although it takes a short transition for neural adaption with the increase of noises, as seen in Figure 5.8d and e, the presented neural network is able to reach a small tracking error no greater than 10×10^{-3} ultimately, which is comparable with the situation without the presence of noises. To further show the improved performance with the presented neural network in the presence of polynomial noises, we additionally consider the track of a square path in the horizontal plane with the size of each edge being 0.3 m. With the same setup of neural controller parameters as the situation to track a circular motion, we run one more set of simulation.

5.6 Summary

In this chapter, we consider the recurrent neural network design for redundancy resolution of manipulators in the presence of polynomial noises. A novel neural control law is presented to tackle this problem. Theorems are established to show the convergence of the neural controller with noisy inputs. Extensive simulations verify the theoretical conclusions. Numerical comparisons with existing neural controllers show significant performance improvements in control accuracy.

6

Using Neural Networks to Avoid Robot Singularity

6.1   Introduction

In recent decades, robotics has attracted considerable attention in scientific research and engineering applications. Much effort has been spent to robotics, and different types of robots have been developed and investigated [61–68]. Among these robots, redundant robot manipulators, possessing more degrees of freedom (DOFs) in joint space than workspace and offering increased control flexibility for complicated tasks, have played a more and more important role in numerous fields of engineering applications [64, 68, 69]. For example, the problem of finite-time stabilization and control of redundant manipulators is investigated in [69], and a controller is designed to attenuate the effects of model nonlinearity, uncertainties, and external disturbances and improve the response characteristics of the system. The forward kinematics of a redundant manipulator provides a nonlinear mapping from its joint space to its operating region in Cartesian space. This nonlinear mapping makes it difficult to directly solve the redundancy resolution problem at the angle level. Instead, in most approaches, the problem is first converted into a problem at the velocity or acceleration level, and solutions are then sought in the converted space. A popular method is to apply the pseudoinverse formulation for obtaining a general solution at the joint-velocity level. However, this strategy suffers from an intensive computational burden because of the need to perform the pseudoinversion of matrices continuously over time, in addition to the weakness that the joint physical limits are usually not considered [70, 71]. Recent progress shows the advantages of unifying the treatment of various constraints of manipulators' redundancy resolution into a quadratic program. Such a quadratic program formulation is general in the sense that it incorporates equality, inequality and bound constraints, simultaneously. For example, reformulated as a quadratic program, the equivalence between different-level redundancy-resolution of redundant manipulators is investigated in [72]. Then, the work is extended in [68] for obstacle avoidance of redundant robot manipulators with the aid of a quadratic program. In addition to the control of redundant manipulators, the quadratic program is also exploited for the other types of robots, e.g. formation control of leader–follower mobile robots' systems in [61, 65, 73].

Neural networks, which feature capabilities of high-speed parallel distributed processing, and can be readily implemented by hardware, have been recognized as a powerful

Kinematic Control of Redundant Robot Arms Using Neural Networks, First Edition.
Shuai Li, Long Jin and Mohammed Aquil Mirza.
© 2019 John Wiley & Sons Ltd. Published 2019 by John Wiley & Sons Ltd.

tool for real-time processing and successfully applied widely in various control systems [64–66, 72, 74–81]. Particularly, using neural networks for the control of robot manipulators has attracted much attention and various related schemes and methods have been proposed and investigated [64–66, 79, 80]. Cai and Zhang present two neural networks for the online solution of a quadratic programming problem existing in the redundancy resolution of manipulators in [72]. They further modify their work in [64] by proposing new noise-tolerant neural networks with application to the motion generation of redundant manipulators. It is worth pointing out here that those neural networks presented in [64, 72] do not consider the bound constraint and thus cannot avoid the joint physical constraints in the manipulator applications. For the direct consideration of inequalities in constrained quadratic programming problems, researchers have attempted to consider the problem in its dual space. Various dual neural networks with different architectures have been proposed [68, 82–85]. The dual neural network based methods are highly efficient for real-time processing and have been successfully used in various applications, including redundant manipulator control.

The performance index plays an important role in quadratic programming formulation based manipulator control, which, to some extent, determines the application potential of redundant manipulators in different industry fields. Therefore, how to model an efficient performance index is an important issue for the redundancy resolution. It is worth pointing out that when a manipulator is at a kinematic singularity configuration, its Jacobian matrix becomes ill-conditioned and rank-deficient [83]. In addition, getting close to a singularity point of the kinematic mapping is also undesirable and unacceptable due to the fact that, in such a state, when the end-effector moves in certain directions, joint velocities and accelerations can be arbitrarily large and this may damage the manipulator. Therefore, research on maximizing the manipulability of manipulators has explicit significance and has been investigated widely. Yoshikawa establishes the first quantitative measure of the manipulability for redundant manipulators at the joint-velocity level in [86]. The seminal study presented by Yoshikawa exploits the pseudoinverse-type formulation to avoid the singularity of redundant manipulators by maximizing the manipulability measure. A multiobjective optimization of a hybrid robotic machine tool is presented in [87], which involves different types of manipulability. Note that most of the aforementioned techniques are based on pseudoinverse-type formulations and do not consider physical limits of manipulators. Zhang et al. present a quadratic programming formulation based manipulability-maximizing scheme, which can handle the physical limits. However, this scheme suffers from an intensive computational burden because of the need to perform the matrix inversion continuously over time in performance index [83]. Moreover, the tradeoff between manipulability and energy consumption is fixed in this scheme, which leads to less adaptation to environments. This chapter makes progress on this front through the modeling of the manipulability optimization of redundant manipulators and the proposal of dynamic neural network to remedy the weaknesses of existing schemes. Note that the scheme presented in [83], together with the other schemes mentioned, will be compared in detail in Section 6.6 to clarify the differences and advantages of the new scheme proposed.

6.2 Preliminaries

In this section, we present useful definitions and convergence lemmas that play an important role in the theoretical derivation in this chapter.

Projection operator. The projection operator for a set $S \subset \mathbb{R}^k$ and $x \in S$ is defined by

$$P_S(x) = \text{argmin}_{y \in S} \| y - x \|^2, \tag{6.1}$$

where $\| \cdot \|$ denotes the Euclidean norm.

Normal cone. For a convex set $S \subset \mathbb{R}^k$ and $x \in S$, the normal cone of S at x, denoted by $N_S(x)$, is defined as

$$N_S(x) = \{\varphi \in \mathbb{R}^k, (y - x)^T \varphi \leq 0, \forall y \in S\}. \tag{6.2}$$

Monotone mapping. A mapping $F(\cdot)$ is called monotone if for each pair of points x, y, there is

$$(x - y)^T(F(x) - F(y)) \geq 0. \tag{6.3}$$

The above definition generalizes the definition of monotonicity from univariable mappings to multivariable mappings. Note that it is possible to verify the monotonicity of a univariable mapping by checking whether its derivative is greater than zero. Similarly, this property can also be extended to the general multivariable mapping $F(\cdot)$. For a continuously differentiable mapping $F(\cdot)$, it is a monotone mapping if

$$\nabla F + \nabla^T F \geq 0, \tag{6.4}$$

where ≥ 0 means the left-hand side of this operator is a positive semidefinite matrix.

For the projection operator and the normal cone, their relationship is summarized in the following lemma.

Lemma 6.1 (Projection and normal cone [59]) For a compact convex set $S \in \mathbb{R}^k$ and $x \in S$, $z \in N_S(x)$ is equivalent to

$$P_S(x + z) = x. \tag{6.5}$$

For a general dynamic systems, the so-called LaSalle's invariance principle applies, as stated below.

Lemma 6.2 (LaSalle's invariance principle [88]) Let S be a compact set that is positively invariant with respect to the dynamics of a system $\dot{x} = g(x)$. Let $V = V(x) \in \mathbb{R}$ be a continuously differentiable function such that $\dot{V} \leq 0$ in S. Let E be the set of all points in S where $\dot{V} = 0$. Let M be the largest invariant set in E. Then every solution starting in S approaches M as $t \to \infty$.

For a time-varying function, Barbalat's lemma holds.

Lemma 6.3 (Barbalat's lemma [88]) If $f(t) \in \mathbb{R}^k$ has a finite limit as $t \to \infty$ and if $\dot{f}$ is uniformly continuous (or $\ddot{f}$ is bounded), then $\lim_{t \to \infty} \dot{f}(t) = 0$.

For a class of well studied dual neural networks, one sufficient condition for its convergence has been established as in the following.

Lemma 6.4 (Convergence for a class of neural networks [57]) Assume that $F(x)$ is monotone and continuously differentiable. The following dynamic system

$$\varepsilon \dot{x} = -x + P_S(x - \rho F(x)), \tag{6.6}$$

where $\varepsilon > 0$ and $\rho > 0$ are both positive constants, S is a closed convex set, $P_S(\cdot)$ denotes the projection operator as defined by (6.1). Then, the trajectory corresponding to (6.6) globally converges to its equilibrium point.

6.3 Problem Formulation

In this section, definitions on manipulator kinematics and manipulability are presented for problem formulation.

6.3.1 Manipulator Kinematics

For a k-DOF redundant manipulator with the joint angle $\theta(t) = [\theta_1(t), ..., \theta_k(t)]^T \in \mathbb{R}^k$, its Cartesian coordinate $r \in \mathbb{R}^m$ in the workspace can be described as a nonlinear mapping:

$$r(t) = f(\theta(t)), \tag{6.7}$$

where the mapping $f(\cdot)$ carries mechanical and geometrical information of a manipulator. By definition, the joint space dimension k of a redundant manipulator is greater than the workspace dimension m, i.e. $k > m$. Computing time derivations on both sides of (6.7), we have

$$\dot{r}(t) = J(\theta(t))\dot{\theta}(t). \tag{6.8}$$

where $J(\theta(t)) = \partial f/\partial \theta \in \mathbb{R}^{m \times k}$ is called the Jacobian matrix of $f(\theta(t))$, and usually is abbreviated as J. The end-effector $r(t)$ of the redundant manipulator is expected to track the desired path $r_d(t)$, i.e. $r(t) \to r_d(t)$ with $\dot{r}(t) \to \dot{r}_d(t)$. It is worth pointing out here that, in the rest of this chapter, the argument t is frequently omitted for presentation convenience, e.g. by writing $r(t)$ as r.

6.3.2 Manipulability

The manipulability of a robot manipulator is a function of the Jacobian matrix as defined below [88]:

$$\mu = \sqrt{\det(JJ^T)} = \sqrt{\mu_1 \mu_2 \cdots \mu_m} \tag{6.9}$$

where $\mu_i \geq 0$ for $i = 1, 2, ..., m$ is the ith largest eigenvalue of JJ^T (note $JJ^T \geq 0$ and thus its eigenvalues are all non-negative). This measure gives an overall scalar description on the gain from joint velocity θ to $\dot{r}$ and measures the amount of singularity. When the robot arm is singular, i.e. rank$J < m$, μ reaches its least value 0. To increase the manipulability and avoid singularity, a large value of μ is preferable in operation.

6.3.3 Optimization Problem Formulation

Therefore, based on the above analysis, the redundancy resolution of a redundant manipulator with manipulability optimality considered is formulated as

$$\min_{\theta} -\mu, \tag{6.10a}$$

$$r_d = f(\theta). \tag{6.10b}$$

So far the redundancy resolution has been formulated as a constrained optimization with the manipulator kinematics as an equation constraint and the manipulability as a cost to maximize. In this optimization problem, μ, as a function of $J = J(\theta)$, usually is nonconvex relative to θ. Additionally, the equation constraint is also usually nonlinear. Thus, the solution of θ in (6.10) becomes a challenging problem.

6.4 Reformulation as a Constrained Quadratic Program

Here, by incorporating physical constraints and redefining the objective function, we reformulate such a redundancy resolution problem with optimal manipulability considered into a constrained quadratic programming problem.

6.4.1 Equation Constraint: Speed Level Resolution

Equation (6.7) describes the mapping from the joint space to the workspace in position level and has a strong nonlinearity. As shown in (6.8), the mapping in velocity level can be significantly simplified. Defining $v = \dot{r}$ as the end-effector velocity in the workspace and $w = \dot{\theta}$ as the joint angular velocity in the joint space and abbreviating $J(\theta(t))$ as J, Equation (6.8) can be rewritten as

$$v = Jw. \tag{6.11}$$

This equation describes the kinematic mapping from the joint space to the workspace at velocity level. In comparison with the position level mapping (6.7) expressed as a general nonlinear function, the velocity level mapping is much simpler as it is in an affine form. Simply enforcing $\dot{r}_d = Jw$ usually is able to obtain w satisfying the speed requirement $v = \dot{r}_d$, but may suffer from drifting in position due to the loss of explicit information on r_d. Instead, we restrict the motion of the redundant manipulator in the following way to reach a desired position.

$$Jw = -k_0(f(\theta) - r_d) + \dot{r}_d, \tag{6.12}$$

where $k_0 > 0$ is the position error feedback coefficient. Note that $Jw - \dot{r}_d = d(f(\theta) - r_d)/dt$. Accordingly, the above equation can be written as $d(f(\theta) - r_d)/dt = -k_0(f(\theta) - r_d)$ and is able to drive $f(\theta)$ to r_d over time. Following the above reasoning, we can define an equation constraint as follows to represent the speed requirement:

$$Jw = v_d \text{ with } v_d = -k_0(f(\theta) - r_d) + \dot{r}_d. \tag{6.13}$$

6.4.2 Redefinition of the Objective Function

As to $\mu^2/2$, we examine its value in time derivative as

$$\frac{d(\mu^2/2)}{dt} = \frac{d[\det(JJ^T)/2]}{dt}$$

$$= \frac{\det(JJ^T)\operatorname{tr}((JJ^T)^{-1}(J\dot{J}^T + \dot{J}J^T))}{2}, \tag{6.14}$$

where $\operatorname{tr}(\cdot)$ denotes the trace of a matrix. In addition, we have $\operatorname{tr}((JJ^T)^{-1}(J\dot{J}^T)) = \operatorname{tr}((JJ^T)^{-1}J\dot{J}^T) = \operatorname{tr}(\dot{J}J^T(JJ^T)^{-1})$ and $\operatorname{tr}((JJ^T)^{-1}(\dot{J}J^T)) = \operatorname{tr}((\dot{J}J^T)(JJ^T)^{-1})$. Thus, Equation (6.14) can be simplified to

$$\frac{d(\mu^2/2)}{dt} = \frac{d(\det(JJ^T)/2)}{dt}$$

$$= \det(JJ^T)\operatorname{tr}(\dot{J}J^T(JJ^T)^{-1}). \tag{6.15}$$

Note that

$$\dot{\mu} = \frac{1}{\mu}\frac{d(\mu^2/2)}{dt} = \sqrt{\det(JJ^T)}\operatorname{tr}(\dot{J}J^T(JJ^T)^{-1}). \tag{6.16}$$

Overall, the quantity $\dot{\mu}$ describes the change of μ with time. By maximizing its value, the system is enforced to evolve along the direction to increase μ gradually under the constraints. To compute $\dot{J}$, we first define $H_i \in \mathbb{R}^{m \times k}$ for $i = 1, 2, ..., k$ as follows:

$$H_i = \frac{\partial J}{\partial \theta_i} \tag{6.17}$$

with which, $\dot{J}$ is expressed as

$$\dot{J} = \sum_{i=1}^{k} \frac{\partial J}{\partial \theta_i}\dot{\theta}_i = \sum_{i=1}^{k} H_i w_i. \tag{6.18}$$

Incorporating this expression, Equation (6.16) becomes

$$\dot{\mu} = \sqrt{\det(JJ^T)}\operatorname{tr}\left(\left(\sum_{i=1}^{k} H_i w_i\right) J^T(JJ^T)^{-1}\right)$$

$$= \sqrt{\det(JJ^T)} \sum_{i=1}^{k} w_i \operatorname{tr}(H_i J^T(JJ^T)^{-1}). \tag{6.19}$$

For the convenience of later treatment of matrix $(JJ^T)^{-1}$, we now convert the expression of $\operatorname{tr}(H_i J^T(JJ^T)^{-1})$ into a form using $\operatorname{vec}((JJ^T)^{-1})$, which is the vectorization of the matrix $(JJ^T)^{-1}$ formed by stacking the columns of $(JJ^T)^{-1}$ into a single column vector. We have

$$\operatorname{tr}(H_i J^T(JJ^T)^{-1}) = \operatorname{tr}((JH_i^T)^T(JJ^T)^{-1})$$

$$= \operatorname{vec}^T(JH_i^T)\operatorname{vec}((JJ^T)^{-1}). \tag{6.20}$$

With this, Equation (6.19) further becomes

$$\dot{\mu} = \sqrt{\det(JJ^T)} \sum_{i=1}^{k} w_i \operatorname{vec}^T(JH_i^T)\operatorname{vec}((JJ^T)^{-1}). \tag{6.21}$$

Note that

$$\sum_{i=1}^{k} w_i \text{vec}^\mathrm{T}(JH_i^\mathrm{T}) = [w_1, w_2, ..., w_k] \begin{bmatrix} \text{vec}^\mathrm{T}(JH_1^\mathrm{T}) \\ \text{vec}^\mathrm{T}(JH_2^\mathrm{T}) \\ ... \\ \text{vec}^\mathrm{T}(JH_k^\mathrm{T}) \end{bmatrix}. \tag{6.22}$$

To simplify the notation, we define a new operator $\Diamond$ to capture the operation in (6.22) on matrices $H_1, H_2, ..., H_k$ and J as below

$$J\Diamond\{H_1, H_2, ..., H_k\} = \begin{bmatrix} \text{vec}^\mathrm{T}(JH_1^\mathrm{T}) \\ \text{vec}^\mathrm{T}(JH_2^\mathrm{T}) \\ ... \\ \text{vec}^\mathrm{T}(JH_k^\mathrm{T}) \end{bmatrix} = \begin{bmatrix} \text{vec}^\mathrm{T}(H_1^\mathrm{T})(I_m \otimes J^\mathrm{T}) \\ \text{vec}^\mathrm{T}(H_2^\mathrm{T})(I_m \otimes J^\mathrm{T}) \\ ... \\ \text{vec}^\mathrm{T}(H_k^\mathrm{T})(I_m \otimes J^\mathrm{T}) \end{bmatrix}$$

$$= \begin{bmatrix} \text{vec}^\mathrm{T}(H_1^\mathrm{T}) \\ \text{vec}^\mathrm{T}(H_2^\mathrm{T}) \\ ... \\ \text{vec}^\mathrm{T}(H_k^\mathrm{T}) \end{bmatrix} (I_m \otimes J^\mathrm{T}), \tag{6.23}$$

where the second equality is obtained as follows: recall the equation $(X_2^\mathrm{T} \otimes X_1)\text{vec}(X_3) = \text{vec}(X_1 X_3 X_2)$ holds for matrices X_1, X_2 and X_3 of proper sizes. Setting $X_1 = J$, $X_2 = I_m$ and $X_3 = H_i^\mathrm{T}$ for $i = 1, 2, ..., k$ in this equation with I_m being the $m \times m$ identity matrix, we thus have $\text{vec}(JH_i^\mathrm{T}) = (I_m \otimes J)\text{vec}(H_i^\mathrm{T})$. In addition, recall the transposition rule over Kronecker product as $(X_1 \otimes X_2)^\mathrm{T} = X_1^\mathrm{T} \otimes X_2^\mathrm{T}$, with which we conclude $\text{vec}^\mathrm{T}(JH_i^\mathrm{T}) = \text{vec}^\mathrm{T}(H_i^\mathrm{T})(I_m \otimes J^\mathrm{T})$.

With the notation defined in (6.23) and the relation (6.22), Equation (6.21) is converted to

$$\dot\mu = \sqrt{\det(JJ^\mathrm{T})}w^\mathrm{T}(J\Diamond\{H_1, H_2, ..., H_k\})\text{vec}((JJ^\mathrm{T})^{-1}). \tag{6.24}$$

The above expression involves the matrix inversion $(JJ^\mathrm{T})^{-1}$, which is usually time-consuming for online computation. In order to avoid this operation, we treat this quantity as a variable and embed it into the solution procedure. To proceed, we re-examine the definition of matrix inverse, which gives the following for $(JJ^\mathrm{T})^{-1}$:

$$JJ^\mathrm{T}(JJ^\mathrm{T})^{-1} = I_m. \tag{6.25}$$

Setting $X_1 = JJ^\mathrm{T}$, $X_2 = I_m$, $X_3 = A$, equality (6.25) is converted to

$$\text{vec}(I_m) = \text{vec}(JJ^\mathrm{T}A) = (I_m \otimes JJ^\mathrm{T})\text{vec}((JJ^\mathrm{T})^{-1}). \tag{6.26}$$

6.4.3 Set Constraint

Due to physical constraints, the angular velocity of a manipulator cannot exceed certain limits, expressed as

$$\underline{w} \le w \le \overline{w}, \tag{6.27}$$

where $\underline{w} \in \mathbb{R}^k$ and $\overline{w} \in \mathbb{R}^k$ are the lower and the upper bounds of w, respectively. The bound expression (6.27) can also be expressed as a convex set constraint:

$$w = \dot\theta \in \Omega, \tag{6.28}$$

where the convex set Ω is defined as

$$\Omega = \{w \in \mathbb{R}^k, \underline{w} \leq w \leq \overline{w}\}. \tag{6.29}$$

For a redundant manipulator with kinematics (6.12) and a given desired workspace velocity v_{d}, the goal of kinematic redundancy resolution is to find a real-time control law $u = h(w, v_{\mathrm{d}})$ such that the workspace velocity error $v - v_{\mathrm{d}}$ converges to zero, and the joint angular velocity converges to the convex constraint set Ω as in (6.28).

6.4.4 Reformulation and Convexification

Define a new variable $\psi \in \mathbb{R}^{m^2}$, as an estimation of $\mathrm{vec}((JJ^{\mathrm{T}})^{-1})$, to replace the expression of $\mathrm{vec}((JJ^{\mathrm{T}})^{-1})$ in Equations (6.24) and (6.26). Then, the manipulability optimization in speed level is formulated as

$$\min_{w,\psi} -\dot{\mu} = -\sqrt{\det(JJ^{\mathrm{T}})}w^{\mathrm{T}}(J\Diamond\{H_1, H_2, ..., H_k\})\psi, \tag{6.30a}$$

$$Jw = v_{\mathrm{d}}, \tag{6.30b}$$

$$(I_m \otimes JJ^{\mathrm{T}})\psi = \mathrm{vec}(I_m), \tag{6.30c}$$

$$w \in \Omega. \tag{6.30d}$$

It is noteworthy that the term $\sqrt{\det(JJ^{\mathrm{T}})}$ is non-negative and independent of the decision variables w and ψ. Therefore, it is equivalent to change the objective function to $w^{\mathrm{T}}(J\Diamond\{H_1, H_2, ..., H_k\})\psi$. In addition, note that the objective function in (6.30) is bilinear but is nonconvex relative to the decision variables. We incorporate one extra term $\| w\|^2$ to regulate the kinematic energy consumption. Additionally, we augment the objective function with the equality constraints (6.30b) and (6.30c) to convexify the objective function as

$$- c_3 w^{\mathrm{T}}(J\Diamond\{H_1, H_2, ..., H_k\})\psi + c_0 \| Jw - v_{\mathrm{d}}\|^2/2$$
$$+ c_1 \| (I_m \otimes JJ^{\mathrm{T}})\psi - \mathrm{vec}(I_m)\|^2/2 + c_2 \| w\|^2/2, \tag{6.31}$$

where $c_0 > 0$, $c_1 > 0$, $c_2 > 0$, and $c_3 > 0$ are constants. Note that this new objective function is quadratic relative to w and ψ with $w^{\mathrm{T}}(J\Diamond\{H_1, H_2, ..., H_k\})\psi$ being a cross term. Now, the reformulated optimization problem can be summarized as

$$\min_{w,\psi} -c_3 w^{\mathrm{T}}(J\Diamond\{H_1, H_2, ..., H_k\})\psi + c_0 \| Jw - v_{\mathrm{d}}\|^2/2$$
$$+ c_1 \| (I_m \otimes JJ^{\mathrm{T}})\psi - \mathrm{vec}(I_m)\|^2/2 + c_2 \| w\|^2/2, \tag{6.32a}$$

$$Jw = v_{\mathrm{d}}, \tag{6.32b}$$

$$(I_m \otimes JJ^{\mathrm{T}})\psi = \mathrm{vec}(I_m), \tag{6.32c}$$

$$w \in \Omega. \tag{6.32d}$$

So far, the redundancy resolution for manipulability optimization has been formulated as a constrained quadratic programming problem. However, the solution to the optimization problem (6.30) cannot be obtained directly.

6.5 Neural Networks for Redundancy Resolution

In this section, we consider developing neural dynamics for real-time solution of the optimal manipulability redundancy resolution problem (6.32). We first convert the problem into the solution of a nonlinear equation set and then establish neural dynamics for solving this nonlinear equation set.

6.5.1 Conversion to a Nonlinear Equation Set

To solve (6.32), we first convert it into a set of nonlinear equations. Define a Lagrange function as follows:

$$L(w \in \Omega, \psi, \lambda, \sigma) = -c_3 w^{\mathrm{T}}(J\Diamond\{H_1, H_2, ..., H_k\})\psi$$
$$+ c_0 \| Jw - v_{\mathrm{d}} \|^2/2 + c_1 \| (I_m \otimes JJ^{\mathrm{T}})\psi - \mathrm{vec}(I_m) \|^2/2$$
$$+ c_2 \| w \|^2/2 + \lambda^{\mathrm{T}}(Jw - v_{\mathrm{d}})$$
$$+ \sigma^{\mathrm{T}}((I_m \otimes JJ^{\mathrm{T}})\psi - \mathrm{vec}(I_m)), \tag{6.33}$$

According to the Karush–Kuhn–Tucker condition [59], the solution of (6.32) has to satisfy the following

$$-\frac{\partial L}{\partial w} \in N_\Omega(w), \quad \frac{\partial L}{\partial \psi} = 0, \tag{6.34}$$

where $N_\Omega(w)$ denotes the normal cone of set Ω at w. Equation (6.34) includes normal cone operation $N_\Omega(w)$ in its expression. It can also be equivalently written in the following form with the aid of projection operators:

$$w = P_\Omega(w - \frac{\partial L}{\partial w}), \tag{6.35}$$

i.e.

$$w = P_\Omega[c_3(J\Diamond\{H_1, H_2, ..., H_k\})\psi - c_0 J^{\mathrm{T}}(Jw - v_{\mathrm{d}})$$
$$+ (1 - c_2)w - J^{\mathrm{T}}\lambda]. \tag{6.36}$$

Together with $\partial L/\partial \psi = 0$ in (6.34), and the equation constraints in (6.32), the nonlinear equations for the optimization problem (6.32) can be summarized as

$$0 = -w + P_\Omega[c_3(J\Diamond\{H_1, H_2, ..., H_k\})\psi$$
$$- c_0 J^{\mathrm{T}}(Jw - v_{\mathrm{d}}) + (1 - c_2)w - J^{\mathrm{T}}\lambda], \tag{6.37a}$$

$$0 = -c_3(J\Diamond\{H_1, H_2, ..., H_k\})^{\mathrm{T}}w + c_1((I_m \otimes JJ^{\mathrm{T}})$$
$$\cdot [(I_m \otimes JJ^{\mathrm{T}})\psi - \mathrm{vec}(I_m)]) + (I_m \otimes JJ^{\mathrm{T}})\sigma, \tag{6.37b}$$

$$0 = Jw - v_{\mathrm{d}}, \tag{6.37c}$$

$$0 = \mathrm{vec}(I_m) - (I_m \otimes JJ^{\mathrm{T}})\psi. \tag{6.37d}$$

The above derivation can be summarized in the following lemma.

Lemma 6.5 The solution to the problem of manipulator redundancy resolution with maximal manipulability, i.e. the constrained optimization problem (6.32), is identical to the solution to the nonlinear equation set (6.37).

6.5.2 Neural Dynamics for Real-Time Redundancy Resolution

So far, we have converted the manipulability optimization problem into a nonlinear equation one. However, solving (6.37), due to its inherent nonlinearity, is a time-consuming task and usually is hard to implemented in real time. Additionally, notice that the nonlinear equation set (6.37) is in nature a time-varying system and its solution thus also varies with time. How to find the solution of (6.37) efficiently and keep track of it with time is a challenging problem. In this section, we present a dynamic neural network, which is able to address the problem recurrently.

For the nonlinear equation set (6.37), we construct a dynamic neural network, as described by the following ordinary differential equations with the equilibrium point identical to (6.37), to solve the redundancy resolution problem:

$$
\begin{aligned}
\varepsilon \dot{w} = -w + P_\Omega [c_3 (J \Diamond \{H_1, H_2, ..., H_k\}) \psi \\
- c_0 J^{\mathrm{T}} (Jw - v_{\mathrm{d}}) + (1 - c_2) w - J^{\mathrm{T}} \lambda],
\end{aligned}
\tag{6.38a}
$$

$$
\begin{aligned}
\varepsilon \dot{\psi} = -c_3 (J \Diamond \{H_1, H_2, ..., H_k\})^{\mathrm{T}} w + (I_m \otimes JJ^{\mathrm{T}}) \sigma + \\
c_1 ((I_m \otimes JJ^{\mathrm{T}})[(I_m \otimes JJ^{\mathrm{T}}) \psi - \mathrm{vec}(I_m)]),
\end{aligned}
\tag{6.38b}
$$

$$
\varepsilon \dot{\lambda} = Jw - v_{\mathrm{d}},
\tag{6.38c}
$$

$$
\varepsilon \dot{\sigma} = \mathrm{vec}(I_m) - (I_m \otimes JJ^{\mathrm{T}}) \psi,
\tag{6.38d}
$$

where scaling factor $\varepsilon > 0$.

The architecture of the proposed dynamic neural network (6.39) is shown in Figure 6.1 for the situation with $m = 3$, $k = 6$, (m is the dimension of workspace, k is the number of joint angle of a PUMA 560 redundant robot manipulator investigated in the ensuing simulations). From this figure, it is clear that the neural network is organized in a one-layer architecture, which is composed of $k + m + 2m^2$ neurons, and is a nonlinear layer with dynamic feedback. This layer of neurons is associated with the state variables $w \in \mathbb{R}^k$, $\psi \in \mathbb{R}^{m^2}$, $\lambda \in \mathbb{R}^m$, and $\sigma \in \mathbb{R}^{m^2}$, and gets input from v_{d}. It follows (6.39a), (6.39b), (6.39c), and (6.39d) for dynamic updates and maps the state variables to the output, which is the joint velocity w of the robot manipulator.

6.5.3 Convergence Analysis

This section proves stability and convergence on the proposed neural network (6.39) via the following theorem.

Theorem 6.1 The dynamical neural network (6.39) is stable and is globally convergent to the optimal solution of (6.32).

Proof: Note that, in this proof, $J \Diamond H$ stands for $J \Diamond \{H_1, H_2, ..., H_k\}$ due to space limitation. By letting $\Upsilon = [w, \psi, \lambda, \sigma]^{\mathrm{T}}$, dynamical neural network (6.39) can be formulated as

$$
\varepsilon \dot{\Upsilon} = -\Upsilon + P_\Theta [\Upsilon - \rho F(\Upsilon)],
\tag{6.39}
$$

where $\rho = 1$, $P_\Theta(\cdot) = [P_\Omega, P_\Lambda]^{\mathrm{T}} \in \mathbb{R}^{2m^2 + m + k}$ with $P_\Lambda \in \mathbb{R}^{2m^2 + m}$ and with the upper and lower limits of set Λ being $\pm \infty$. Defining $F_1(\Upsilon) = -c_3 (J \Diamond H) \psi + c_0 J^{\mathrm{T}} (Jw - v_{\mathrm{d}}) + c_2 w +$

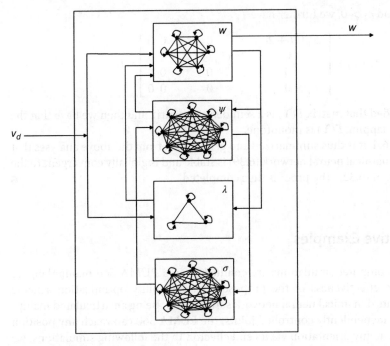

Figure 6.1 Neural network architecture.

$J^T\lambda$ and $F_2(\Upsilon) = c_3(J\Diamond H)^T w - c_1((I_m \otimes JJ^T) \cdot [(I_m \otimes JJ^T)\psi - \text{vec}(I_m)]) - (I_m \otimes JJ^T)\sigma$,

$$F(\Upsilon) = \begin{bmatrix} F_1(\Upsilon) \\ F_2(\Upsilon) \\ v_d - Jw \\ (I_m \otimes JJ^T)\psi - \text{vec}(I_m) \end{bmatrix} \in \mathbb{R}^{k+2m^2+m},$$

and

$$\nabla F(\Upsilon) = \begin{bmatrix} \dfrac{\partial F_1(\Upsilon)}{\partial w} & \dfrac{\partial F_1(\Upsilon)}{\partial \psi} & \dfrac{\partial F_1(\Upsilon)}{\partial \lambda} & \dfrac{\partial F_1(\Upsilon)}{\partial \sigma} \\ \dfrac{\partial F_2(\Upsilon)}{\partial w} & \dfrac{\partial F_2(\Upsilon)}{\partial \psi} & \dfrac{\partial F_2(\Upsilon)}{\partial \lambda} & \dfrac{\partial F_2(\Upsilon)}{\partial \sigma} \\ -J & 0 & 0 & 0 \\ 0 & (I_m \otimes JJ^T) & 0 & 0 \end{bmatrix}$$

$$= \begin{bmatrix} c_0 J^T J + c_2 I & -c_3 J\Diamond H & J^T & 0 \\ c_3(J\Diamond H)^T & c_1(I_m \otimes JJ^T)^2 & 0 & -I_m \otimes JJ^T \\ -J & 0 & 0 & 0 \\ 0 & I_m \otimes JJ^T & 0 & 0 \end{bmatrix}.$$

Therefore, $F(\Upsilon)$ is continuously differentiable in view of the existence of $\nabla F(\Upsilon)$. In addition, it can be readily deduced that $I_m \otimes JJ^T = (I_m \otimes JJ^T)^T$. In view of the facts that

$c_0 > 0$, $c_1 > 0$, and $c_2 > 0$, we further have

$$\nabla F(\Upsilon) + \nabla^T F(\Upsilon) = \begin{bmatrix} 2c_0 J^T J + 2c_2 I & 0 & 0 & 0 \\ 0 & 2c_1 (I_m \otimes JJ^T)^2 & 0 & 0 \\ 0 & 0 & 0 & 0 \\ 0 & 0 & 0 & 0 \end{bmatrix} \geq 0.$$

It can be concluded that matrix $F(\Upsilon)$ is a semidefinite matrix and then we have that the corresponding mapping $F(\Upsilon)$ is monotone.

From Lemma 6.4, it is thus summarized and generalized from the above analyses that the proposed dynamical neural network (6.39) is stable and is globally convergent to the optimal solution of (6.32). The proof is thus completed. ∎

6.6 Illustrative Examples

In this section, computer simulations are conducted on a PUMA 560 manipulator to demonstrate the effectiveness of the proposed manipulability optimization scheme (6.32) as well as its dynamical neural network solver (6.39). Being an articulated manipulator with six independently controlled joints, the PUMA 560 can reach any position in its workspace at any orientation via its end effector. In the following simulations, we consider only the three-dimensional position control of the end-effector, and thus, the PUMA 560 can be deemed as a redundant manipulator for this particular task.

6.6.1 Manipulability Optimization via Self Motion

In this subsection, we perform simulations via self motion of a PUMA 560 manipulator, i.e. by using different schemes to drive the manipulator from one state with low manipulability to another state with high manipulability and without moving the end-effector in Cartesian space. With parameters being set as $\varepsilon = 0.01$, $c_0 = c_1 = 1$, $c_2 = c_3 = 0.01$, $k_0 = 5$, $\Omega = [-2, 2]^6$ and the rest being initially set as 0 (e.g. $\dot{w}$), a typical simulation run from a random initialization is generated as shown in Figures 6.2 and 6.3. Specifically, as illustrated in Figure 6.2a, the manipulator adjusts its state with the end-effector being unmoved. As shown in Figure 6.2b, the manipulability measure $\mu = \sqrt{\det(JJ^T)}$ of the PUMA 560 manipulator increases to 0.17, which means that the proposed manipulability optimization scheme (6.32) is effective. The corresponding simulation results of the detailed joint-angle, joint-velocity and position-error profiles for the manipulability optimization of the PUMA 560 via self motion are illustrated in Figure 6.3. It can be seen from Figure 6.3 that, after a short-time adjustment, the joint angle θ of the PUMA 560 converges to a constant value in Figure 6.3a, and correspondingly the joint velocity $\dot{\theta}$ approaches zero as shown in Figure 6.3b. The control error $e = r - r_d$, where r_d represents the reference position in workspace, approaches zero with time as shown in Figure 6.3c. Overall, as shown in Figures 6.2 and 6.3, the manipulability optimization of the PUMA 560 manipulator via self motion synthesized by the proposed manipulability optimization scheme (6.32) as well as its dynamical neural network solver (6.39) are illustrated.

Figure 6.2 Simulation results on (a) motion trajectories and (b) manipulability measures for the manipulability optimization of PUMA 560 manipulator via self motion with its end-effector fixed at [0.55, 0, 1.3] m in the workspace.

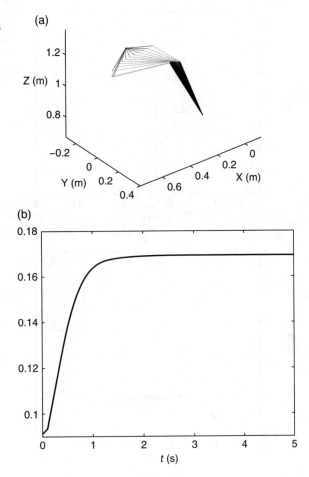

In addition, as a comparison, simulation results of the PUMA 560 manipulator synthesized by the scheme presented in [72] are illustrated in Figure 6.4. With all the parameters being same as those in the simulations shown in Figures 6.2 and 6.3, the PUMA 560 manipulator does not move and its manipulability keeps at the value of 0.091. In summary, these simulation results demonstrate the effectiveness of the proposed manipulability optimization scheme (6.32) in the increase of the manipulability as well as the singularity avoidance.

6.6.2 Manipulability Optimization in Circular Path Tracking

In this section, computer simulations synthesized by the proposed manipulability optimization scheme (6.32) are conducted to track a circular path. Specifically, the reference position of the PUMA 560's end-effector moves at an angular speed of 0.2 rad/s along a circle centered at [0.25, 0, 1.3] m with radius 0.2 m and a revolution angle of

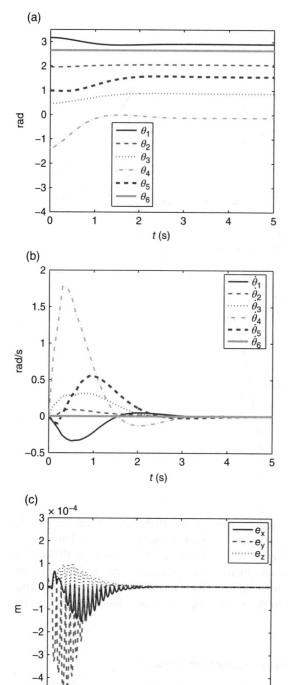

(a)

(b)

(c)

Figure 6.3 Simulation results of (a) joint-angle, (b) joint-velocity and (c) position-error profiles for the manipulability optimization of PUMA 560 manipulator via self motion with its end-effector fixed at [0.55, 0, 1.3] m in the workspace.

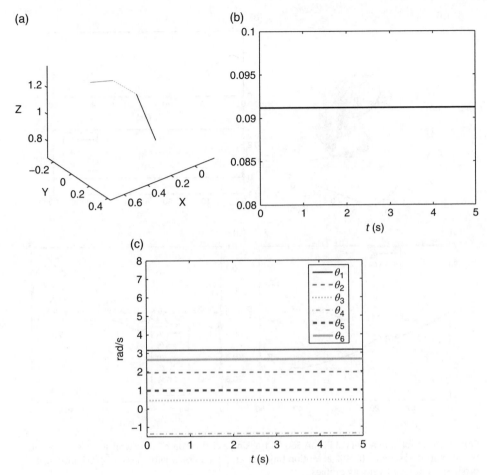

Figure 6.4 Simulation results of (a) motion trajectories, (b) manipulability measures and (c) joint-angle profiles of PUMA 560 synthesized by the scheme presented in [72] with its end-effector fixed at [0.55, 0, 1.3] m in the workspace.

$30°$ around the x axis. In addition, the parameters are set as $\varepsilon = 0.01$, $c_0 = c_1 = 1$, $c_2 = c_3 = 0.01$, $k_0 = 5$, $\Omega = [-0.5, 0.5]^6$ with the rest being initially set as 0 (e.g. $\dot{w}$). A typical simulation run, as shown in Figure 6.5, can be generated using the proposed manipulability optimization scheme (6.32) starting from a random initial configuration. As shown in Figure 6.5a, the end-effector of the PUMA 560 manipulator tracks the circular path successfully in the three-dimensional work-plane with its initial position being not on the desired path. The manipulability measures and position-error profiles are illustrated in Figure 6.5b, from which we can observe that the manipulator is always away from singularity and the position error converges to zero after a short-time transient. In addition, the joint angle θ varies with time in Figure 6.5c. As shown in Figure 6.5d, the joint velocity $\dot{\theta}$ is kept within the limits. These results verify the effectiveness of our method.

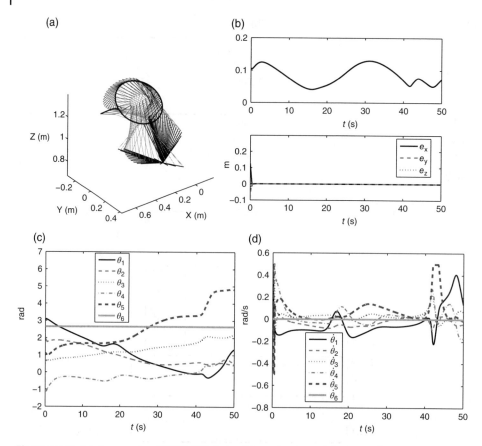

Figure 6.5 Simulation results of PUMA 560 for tracking a circular path in the workspace synthesized by the proposed scheme (6.32). (a) Motion trajectories; (b) manipulability measures; (c) joint-angle profiles; and (d) joint-velocity profiles.

6.6.3 Comparisons

In this section, we compare the proposed manipulability optimization scheme (6.32) with existing solutions for tracking control of redundant manipulators presented in [70, 72, 83–85]. The scheme presented in [72] extends results in [85] from an equivalent relationship viewpoint on the velocity-level redundancy resolution and acceleration-level resolution. In this chapter, we focus on the velocity-level resolution and the proposed manipulability optimization scheme (6.32) that applies to it. The proposed scheme in this chapter is able to deal with the manipulability optimization problem in an inverse-free manner, while existing ones [70, 72, 83–85] cannot handle such a knotty problem. In addition, due to the lack of direct position feedback in [72, 85], the initialization of the manipulator's end-effector is strictly restricted to the desired position on the desired path. Moreover, another major difference is that the proposed scheme (6.32) always maximizes the manipulability of the manipulators, while others show less satisfying performance.

It also can be observed from Figure 6.6 that, for the situation of well-conditioned Jacobian matrix, the total computational time related to the proposed scheme (6.32)

Figure 6.6 Manipulability measures of PUMA 560 by different schemes.

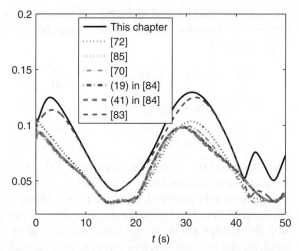

is 2.7 s, which is longer than that of the rest. This is because that extra effort is paid for by the optimization of manipulability. The total computational time of scheme (6.32) is much shorter than the task duration (i.e. 50 s), which means that such a scheme can be employed to the online solution of motion generation of redundant robot manipulators. In addition, it was found in our experiment that, for the situation of almost singular Jacobian matrix, the scheme presented in [83] uses 1029 s to fulfill the task, much longer than the task duration (i.e. 50 s), which means that such a scheme cannot be employed for the online solution of motion generation of redundant robot manipulators. By contrast, the total computational time related to scheme (6.32) is 12.4 s, which is much shorter due to the avoidance of time-consuming matrix inversion when the matrix is almost singular. It is also noteworthy that as the benefit of manipulability optimization, the proposed scheme can quickly solve the control action while others without optimizing manipulability have to update control actions frequently when their Jacobian reaches singular, resulting in the proposed scheme consuming less time than those without manipulability optimization [72, 84, 85]. Moreover, it can be observed in our experiment that, for the situation of singular Jacobian matrix, the scheme presented in [83] cannot fulfill such a given path-tracking task, which is mainly because of the real-time matrix-inversion operation involved in the scheme. Therefore, this scheme fails when encountering a singular Jacobian matrix. By contrast, as shown in our experiment, even without the ability to maximize the manipulability during the task execution process, these quadratic-program-based schemes presented in [[72, 84, 85] could generate an approximated solution with no feasible solution at the cost of large position errors. As a quadratic-program-based control law designed for manipulability optimization of redundant robot manipulators in an inverse-free manner, the proposed manipulability optimization scheme (6.32) could handle such a knotty problem in an acceptable way. That is, our scheme could always maximize the manipulability in the situation of nonsingular Jacobian matrix, and generate an approximated solution for commanding the manipulator motion in the situation of singular Jacobian matrix with superior computational efficiency and lesser position errors compared with other existing schemes.]

Besides, it is worth comparing here the manipulability optimization scheme (6.32) and the manipulability-maximizing scheme presented in [83], both of which are exploited to maximize the manipulability during the motion generation of redundant robot manipulators. As a basis for discussion, the manipulability optimization scheme presented in [83] is rewritten as

$$\min_{w} \quad \frac{1}{2}w^{\mathrm{T}}w - (\det(JJ^{\mathrm{T}})\mathrm{tr}((JJ^{\mathrm{T}})^{-1}\frac{\partial(JJ^{\mathrm{T}})}{\partial\theta}))^{\mathrm{T}}w \tag{6.40a}$$

$$Jw = v_{\mathrm{d}}, \tag{6.40b}$$

$$w \in \Omega. \tag{6.40c}$$

The reasons for the different performance of these schemes can be explained intuitively as follows: the manipulability optimization scheme (6.40) has more computational load due to the matrix inversion involved, i.e. the matrix JJ^{T} has to be inverted in real time. On the contrary, the proposed manipulability optimization scheme (6.32) is able to deal with the manipulability optimization problem in an inverse-free manner. On the one hand, for the well-conditioned Jacobian matrix, such a matrix inversion operation can be readily done and scheme (6.40) seems to be effective in this situation. On the other hand, it is unnecessary to implement a scheme for maximizing the manipulability with well-conditioned Jacobian matrix since the given task is far away from singularity. It is worth noting that for the matrix inversion operation existing in scheme (6.40), it would take much longer than scheme (6.32) to compute the inverse of an almost singular Jacobian matrix, which may violate real-time requirements on the motion generation of redundant robot manipulators. What is worse, for the situation of singular Jacobian matrix, the manipulability optimization scheme (6.40) cannot work since there does not exist the inverse of a singular matrix. Note that as long as one eigenvalue of the matrix JJ^{T} is zero, scheme (6.40) would fail to complete the task. However, the proposed manipulability optimization scheme (6.32) can handle such a knotty problem.

6.6.4 Summary

In this chapter, we have established a dynamic neural network for recurrent calculation of manipulability-maximal control actions for redundant manipulators under physical constraints in an inverse-free manner. By expressing position tracking and matrix inversion as equality constraints, physical limits as inequality constraints, and velocity-level manipulability, which is affine to the joint velocities, as the objective function, the manipulability optimization scheme has been further formulated as a constrained quadratic program. Then, a dynamic neural network with rigorously provable convergence has been constructed to solve such a problem online. Computer simulations show that compared with existing methods the proposed scheme can raise the manipulability by 40% on average, which substantiates the efficacy, accuracy and superiority of the proposed scheme.

Part II

Neural Networks for Parallel Robot Control

7

Neural Network Based Stewart Platform Control

7.1 Introduction

Kinematically redundant manipulators are referred to as those which have more degrees of freedom (DOFs) than required to determine the position and orientation. The redundancy of parallel manipulators can be utilized to overcome the obstacles, singularities [90], increasing workspace, improving dexterity and to optimize the performance and to achieve a smooth end-effector motion task. As redundant robots have more DOFs than required, there usually exist multiple solutions for kinematic control, which motivates us to consider exploiting the extra DOFs to improve the control performance.

The inverse kinematics problem is one of the fundamental tasks in understanding the operability of parallel manipulators, i.e. to find the actuator inputs, providing the desired end-effector trajectories. Conventional design of the parallel mechanisms often encounter singularity problems. The design of redundancy in the parallel mechanism often provides an effective remedy. In [91], the authors proposed a new 3-DOF symmetric spherical 3-UPS/S parallel mechanism with three prismatic actuators, and studied the kinematics, statics, and workspace of the mechanism. In [91], a 2(SP+SPR+SPU) serial-parallel manipulator was considered. Based on the analysis, they designed three new types of kinematically redundant parallel mechanisms, including a new redundant 7-DOF Stewart platform. In [92], the damped least square method was utilized to tackle the singularity problem. However, it only modifies the end-effector path in terms of velocity.

Due to the outstanding performance in parallel processing and recursive computation, dynamic neural networks, as a special type of neural networks, have long been employed as a powerful tool for the control of conventional serial robot arms. Theoretical investigation of new types of neural models, and the applications of them to various real-world problems have also received intensive research attention in recent years. In spite of the great success of dynamic neural networks in both theory and applications, reports are rare on using dynamic neural networks to address the kinematic resolution problem of Stewart platforms. In this chapter, we make progress in this direction. The problem is formulated to a constrained quadratic programming in this chapter. The Karush–Kuhn–Tucker (KKT) conditions of the problem are obtained by considering the problem in its dual space, and then we design a dynamic neural network to solve this problem recurrently. Due to the formulation of the problem from an optimization

Kinematic Control of Redundant Robot Arms Using Neural Networks, First Edition.
Shuai Li, Long Jin and Mohammed Aquil Mirza.
© 2019 John Wiley & Sons Ltd. Published 2019 by John Wiley & Sons Ltd.

perspective, optimality in movement can be directly reached. Actually, for the Stewart platform, its forward kinematics are highly nonlinear and heavily coupled, which impose great challenges for the neural dynamic design to reach the same control goal. Additionally, due to the modeling of constraints as inequalities in the optimization problem, the obtained solution is direct in compliance with the physical constraints. Using convergence theories on projected neural networks, the global convergence of this dynamic neural network to the optimal solution is also proved rigorously. Simulation results verify the effectiveness in the tracking control of the Stewart platform for dynamic motions. The contributions of this study are threefold: first, to the best of our knowledge, this is the first attempt of using dynamic neural networks for kinematic redundancy resolution of parallel Stewart platforms. Secondly, in this chapter, both the minimization of the energy consumption and the feasibility of physical constraints are taken into account. Accordingly, the obtained solution is both optimal and feasible. This is in contrast to other control methodology, which usually cannot guarantee the control action is within a feasible set. Thirdly, different from feedforward neural networks, which usually suffer from local optimality, the dynamic neural approach employed in this chapter is globally optimal.

7.2 Preliminaries

The pose (position and orientation) of a rigid body in three-dimensional space is uniquely determined by a translation, represented by a three-dimensional vector, and a rotation, represented by a 3×3 rotational matrix in terms of three Euler angles. The rotational matrix $\Omega \in \mathbb{R}^{3 \times 3}$ defined by the Euler angles $[\phi_x, \phi_y, \phi_z]^T \in \mathbb{R}^3$ has the following property for its time derivative:

$$\dot{\Omega}\Omega^T = \begin{bmatrix} 0 & -\dot{\phi}_z & \dot{\phi}_y \\ \dot{\phi}_z & 0 & -\dot{\phi}_x \\ -\dot{\phi}_y & \dot{\phi}_x & 0 \end{bmatrix}. \tag{7.1}$$

This property holds for all rotational matrices. Additionally, the rotational matrix is orthogonal, i.e. $\Omega\Omega^T = I$, $\Omega^{-1} = \Omega^T$ with I denoting a 3×3 identity matrix.

For two vectors $u = [u_1, u_2, u_3]^T \in \mathbb{R}^3$ and $v = [v_1, v_2, v_3]^T \in \mathbb{R}^3$, their cross-product, denoted as $\times$, is defined as:

$$\begin{bmatrix} u_1 \\ u_2 \\ u_3 \end{bmatrix} \times \begin{bmatrix} v_1 \\ v_2 \\ v_3 \end{bmatrix} = \begin{bmatrix} u_2 v_3 - u_3 v_2 \\ u_3 v_1 - u_1 v_3 \\ u_1 v_2 - u_2 v_1 \end{bmatrix}. \tag{7.2}$$

The triple product of three vectors in three-dimensional space is defined based on the cross product. For three vectors $u = [u_1, u_2, u_3]^T \in \mathbb{R}^3$, $v = [v_1, v_2, v_3]^T \in \mathbb{R}^3$, and $w = [w_1, w_2, w_3]^T \in \mathbb{R}^3$, their triple product is defined as $(u \times v)^T w$, and equals the following in value

$$(u \times v)^T w = \det \begin{bmatrix} u^T \\ v^T \\ w^T \end{bmatrix}. \tag{7.3}$$

The triple product is invariant under circular shifting:

$$(u \times v)^T w = (v \times w)^T u = (w \times u)^T v. \tag{7.4}$$

The matrix on the right-hand side in Equation (7.1) is a 3×3 skew-symmetric matrix. A general 3×3 skew-symmetric matrix, as can be simply verified, always holds for any $x, y, z \in \mathbb{R}$, $\alpha = [\alpha_1, \alpha_2, \alpha_3]^T \in \mathbb{R}^3$,

$$\begin{bmatrix} 0 & -z & y \\ z & 0 & -x \\ -y & x & 0 \end{bmatrix} \alpha = \begin{bmatrix} x \\ y \\ z \end{bmatrix} \times \alpha. \tag{7.5}$$

Due to the above relation, it is common to use $[x, y, z]^T_\times$ to represent a three-dimensional skew-symmetric matrix as:

$$\begin{bmatrix} 0 & -z & y \\ z & 0 & -x \\ -y & x & 0 \end{bmatrix} \alpha = \begin{bmatrix} x \\ y \\ z \end{bmatrix}_\times. \tag{7.6}$$

7.3 Robot Kinematics

The Stewart platform is a typical parallel mechanism and can be extended to different forms by modifying its mechanisms. It includes a mobile platform on the top, a fixed base, and six independent driving legs connecting the aforementioned two parts. The two ends of each leg are fixed on the mobile platform and the fixed based, respectively, using universal joints. Each leg can be actuated to change its length by the adjustment of the distance between the two fixed points on the platform and the base. All together, the six legs collaborate to adjust the orientation and position of the mobile platform by changing their lengths.

7.3.1 Geometric Relation

For the Stewart platform, the global coordinate is fixed on the base and the platform coordinate is fixed on the mobile platform. $a_i \in \mathbb{R}^3$ for $i = 1, 2, ..., 6$ represents the position in global coordinates of the ith connection point on the base. $b_i' \in \mathbb{R}^3$ for $i = 1, 2, ..., 6$ represents the position in platform coordinates of the ith connection point on the platform. We use b_i to represents its position in the global coordinate. $d_i = b_i - a_i$ for $i = 1, 2, ..., 6$ represents the vector corresponding to the ith leg, which points from the base to the platform. For a point $x' \in \mathbb{R}^3$ in the platform coordinate, its position $x \in \mathbb{R}^3$ in the global coordinate is obtained after a rotational and translational transformation:

$$x = p + Qx', \tag{7.7}$$

where $p = [x_p, y_p, z_p]^T \in \mathbb{R}^3$ is the global coordinate of the zero position in the platform coordinate, and it corresponds to the translational transformation,

$Q \in \mathbb{R}^{3\times3}$ is the rotational matrix, which is uniquely defined by the Euler angles $\theta = [\theta_x, \theta_y, \theta_z]^T \in \mathbb{R}^3$

$$Q = Q_z Q_y Q_x$$

$$Q_x = \begin{bmatrix} 1 & 0 & 0 \\ 0 & \cos\theta_x & \sin\theta_x \\ 0 & -\sin\theta_x & \cos\theta_x \end{bmatrix}$$

$$Q_y = \begin{bmatrix} \cos\theta_y & 0 & -\sin\theta_y \\ 0 & 1 & 0 \\ \sin\theta_y & 0 & \cos\theta_y \end{bmatrix}$$

$$Q_z = \begin{bmatrix} \cos\theta_z & \sin\theta_z & 0 \\ -\sin\theta_y & \cos\theta_y & 0 \\ 0 & 0 & 1 \end{bmatrix}. \tag{7.8}$$

Following Equation (7.7), as to the ith connection point on the platform, i.e. the ones with $x = b_i$ in the global coordinates or the ones with $x' = b'_i$ in the platform coordinates, we have

$$b_i = p + Qb'_i. \tag{7.9}$$

Therefore, the ith leg vector can be further expressed as

$$d_i = b_i - a_i = p + Qb'_i - a_i. \tag{7.10}$$

For the vector d_i, we define $r_i = \| d_i \|$ to represent its length. Accordingly, we have

$$r_i = \| p + Qb'_i - a_i \|. \tag{7.11}$$

Notice that both a_i and b_i are constants and are determined by the geometric structure. $p = [x_p, y_p, z_p]^T$ defines the translation of the platform, and Q, as a function of the Euler angles $\theta = [\theta_x, \theta_y, \theta_z]^T$, defines the rotation of the platform. Overall, the right-hand side of (7.11) depends on the pose variables of the platform $\pi = [x_p, y_p, z_p, \theta_x, \theta_y, \theta_z]^T \in \mathbb{R}^6$ while the left-hand side of (7.11) is the length of the leg, which is controlled for actuation. In this sense, Equation (7.11) for $i = 1, 2, .., 6$ defines the kinematic relation between the actuation variables and the pose variables. For a six-dimensional reference pose, the desired leg length r_i can be directly obtained from (7.11). However, in real applications, the reference is usually not six-dimensional. For example, for surgical applications of the Stewart platform, people may only care about the position of an end-effector on the platform, instead of its orientation. In this situation, the reference is three-dimensional and we have three additional DOFs as redundancy. For such a situation, we usually have an infinite number of feasible solutions of r_i for $i = 1, 2, ..., 6$ to reach the reference. Among the feasible solutions, we may be able to identify one, which outperforms others in terms of certain optimization criteria. This intuitive analysis motivates us to model it as an optimization problem and identify the optimal one for improved performance. However, due to the nonlinearity of Equation (7.11), direct treatment of Equation (7.11) is technically prohibitive. Instead of a direct solution in position space, we turn to solve the problem in its velocity space to exploit the approximate linearity.

7.3.2 Velocity Space Resolution

For easy treatment, Equation (7.11) is rewritten as

$$r_i^2 = (p + Qb_i' - a_i)^T(p + Qb_i' - a_i). \tag{7.12}$$

To obtain the velocity space relations, we first compute the time derivative on both sides of (7.12), which yields

$$
\begin{aligned}
r_i \dot{r}_i &= (p + Qb_i' - a_i)^T(\dot{p} + \dot{Q}b_i' + Q\dot{b}_i' - \dot{a}_i) \\
&= (p + Qb_i' - a_i)^T(\dot{p} + \dot{Q}b_i').
\end{aligned} \tag{7.13}
$$

Recall that both a_i and b_i' are constants and their time derivatives, $\dot{a}_i$ and $\dot{b}_i'$, equal to zero. For the rotational matrix Q, according to the preliminary equations (7.1) and (7.6), it has the following property for its time derivative

$$
\dot{Q}Q^T =
\begin{bmatrix}
0 & -\dot{\theta}_z & \dot{\theta}_y \\
\dot{\theta}_z & 0 & -\dot{\theta}_x \\
-\dot{\theta}_y & \dot{\theta}_x & 0
\end{bmatrix}
=
\begin{bmatrix}
\dot{\theta}_x \\
\dot{\theta}_y \\
\dot{\theta}_z
\end{bmatrix}_\times
= \dot{\theta}_\times. \tag{7.14}
$$

Therefore, $\dot{Q}$ can be written as follows:

$$\dot{Q} = \dot{\theta}_\times (Q^T)^{-1} = \dot{\theta}_\times Q. \tag{7.15}$$

Substituting (7.15) into (7.13) yields

$$
\begin{aligned}
r_i \dot{r}_i &= (p + Qb_i' - a_i)^T(\dot{p} + \dot{\theta}_\times Qb_i') \\
&= d_i^T(\dot{p} + \dot{\theta}_\times Qb_i') \\
&= d_i^T \dot{p} + \left((\dot{\theta} \times (Qb_i')) \right)^T d_i \\
&= d_i^T \dot{p} + \left((Qb_i') \times d_i \right)^T \dot{\theta} \\
&= \begin{bmatrix} d_i^T & \left((Qb_i') \times d_i \right)^T \end{bmatrix} \begin{bmatrix} \dot{p} \\ \dot{\theta} \end{bmatrix}.
\end{aligned} \tag{7.16}
$$

In the above equation, Equations (7.10) and (7.4) are used for the derivation in the second line and the derivation in the penultimate line, respectively. Noticing that $r_i = \lVert d_i \rVert > 0$ could be guaranteed by the mechanical structure, we have the following result

$$
\begin{aligned}
\dot{r}_i &= \frac{1}{r_i} \begin{bmatrix} d_i^T & \left((Qb_i') \times d_i \right)^T \end{bmatrix} \begin{bmatrix} \dot{p} \\ \dot{\theta} \end{bmatrix} \\
&= \frac{1}{r_i} \begin{bmatrix} d_i^T & \left((Qb_i') \times d_i \right)^T \end{bmatrix} \dot{\pi}.
\end{aligned} \tag{7.17}
$$

For the six-dimensional vector $r = [r_1, r_2, ..., r_6]^T$, we have the compact matrix form as follows:

$$\dot{r} = A_1 \dot{\pi}, \tag{7.18}$$

where

$$
A_1 = \begin{bmatrix} \frac{1}{r_1}d_1^T & \frac{1}{r_1}\left((Qb_1') \times d_1\right)^T \\ \frac{1}{r_2}d_2^T & \frac{1}{r_2}\left((Qb_2') \times d_2\right)^T \\ \cdots & \cdots \\ \frac{1}{r_6}d_6^T & \frac{1}{r_6}\left((Qb_6') \times d_6\right)^T \end{bmatrix}
$$

$$
= \begin{bmatrix} \frac{1}{r_1}(p + Qb_1' - a_1)^T & \frac{1}{r_1}\left((Qb_1') \times (p - a_1)\right)^T \\ \frac{1}{r_2}(p + Qb_2' - a_2)^T & \frac{1}{r_2}\left((Qb_2') \times (p - a_2)\right)^T \\ \cdots & \cdots \\ \frac{1}{r_6}(p + Qb_6' - a_6)^T & \frac{1}{r_6}\left((Qb_6') \times (p - a_6)\right)^T \end{bmatrix}. \tag{7.19}
$$

Equation (7.18) gives the kinematic relation of a 6-DOF Stewart platform from the velocity of the pose variables to the speed of the legs.

7.4 Problem Formulation as Constrained Optimization

Compared with Equation (7.11), Equation (7.18) significantly simplifies the problem as $\dot{r}$ in (7.18) is now affine to the $\dot{\pi}$), while the relation between r_i and π (or p and Q) in Equation (7.11) is nonlinear, or even nonconvex to the pose variables. Similar to our analysis before, in the case that the reference pose velocity is given in six dimensions, the solution of $\dot{r}$ can be solved directly from (7.18). However, in the situation that the reference pose velocity is described in dimensions lower than six, the extra redundancies are available to reach improved performance. The following equality models the reference velocity constraint in reduced dimensions

$$
\alpha = A_2 \dot{\pi}, \tag{7.20}
$$

where the reference vector $\alpha \in \mathbb{R}^m$ with $0 < m < 6$ is pre-given, and the matrix $A_2 \in \mathbb{R}^{m \times 6}$ is the transformation matrix and is also pre-given. As an example, if we would like to maintain the platform at a given height, i.e. $\dot{p}_z = 0$, we set $\alpha = 0$ with $m = 1$ and $A_2 = [0, 0, 1, 0, 0, 0]^T$ in (7.20). Due to the extra design freedom, the value of $\dot{\pi}$ usually cannot be uniquely solved from (7.20). We thus define the following criteria to optimize the solution

$$
\min_{(\dot{\pi}, \tau)} \frac{1}{2}\dot{\pi}^T \Lambda_1 \dot{\pi} + \frac{1}{2}\tau^T \Lambda_2 \tau, \tag{7.21}
$$

where $\Lambda_1 \in \mathbb{R}^{6 \times 6}$ and $\Lambda_2 \in \mathbb{R}^{6 \times 6}$ are both symmetric constant matrices and are both positive definite, and $\tau = \dot{r}$ is the controllable speed of the platform legs. In application, the term $\dot{\pi}^T \Lambda_1 \dot{\pi}$ can be used to specify the kinematic energy (including translational kinetic energy and rotational kinetic energy) by choosing a proper weighting matrix Λ_1 (say choosing Λ_1 as one formed from the mass of the platform and its moment of inertia for the kinematic energy case), and the term $\tau^T \Lambda_2 \tau$ is the input power consumed by the robotic system. This objective function also follows the convention of control theory. The decision variable τ, which is controlled by the actuators, is subjected to physical constraints. In this chapter, we model the physical constraints as linear inequalities in the following form:

$$B\tau \leq b, \tag{7.22}$$

where $B \in \mathbb{R}^{k \times 6}$ and $b \in \mathbb{R}^k$ with k being an integer. Note that constraints are not imposed on the variable $\dot{\pi}$ since its value usually is specified as a feasible one in the planning stage. In summary, with Equation (7.21) as the object function, Equation (7.18) as the mapping relation, (7.20) for the reference tracking requirements, and (7.22) as physical constraints, we can formulate the kinematic control problem of the Stewart platform as the following constrained programming:

$$\min_{(\dot{\pi},\tau)} \quad \frac{1}{2}\dot{\pi}^T \Lambda_1 \dot{\pi} + \frac{1}{2}\tau^T \Lambda_2 \tau \tag{7.23a}$$

$$\text{s.t.} \quad \tau = A_1 \dot{\pi} \tag{7.23b}$$

$$\alpha = A_2 \dot{\pi} \tag{7.23c}$$

$$B\tau \leq b. \tag{7.23d}$$

Due to the presence of both equation and inequality constraints in the optimization problem (7.23), usually it cannot be solved analytically. Conventional approaches introduce extra penalty terms formed by the constraints to the objective function and solve the problem numerically using gradient descent along the new objective function. However, penalty-based approaches only reach an approximate solution of the problem and thus are not suitable for error-sensitive applications. Instead of using this approximate approach, in the next section, we will propose a dynamic neural network solution, which asymptotically converges to the theoretical solution.

7.5 Dynamic Neural Network Model

In this section, we first consider the optimization problem (7.23) in its dual space and then present a neural network to solve it dynamically. After that, we investigate the hardware realization of the proposed model.

7.5.1 Neural Network Design

According to the KKT conditions, the solution of problem (7.23) satisfies:

$$\Lambda_1 \dot{\pi} - A_1^T \lambda_1 - A_2^T \lambda_2 = 0 \tag{7.24a}$$

$$\Lambda_2 \tau + \lambda_1 + B^T \mu = 0 \tag{7.24b}$$

$$\tau = A_1 \dot{\pi} \tag{7.24c}$$

$$\alpha = A_2 \dot{\pi} \tag{7.24d}$$

$$\begin{cases} \mu > 0 & \text{if } B\tau = b \\ \mu = 0 & \text{if } B\tau < b \end{cases} \tag{7.24e}$$

where $\lambda_1 \in \mathbb{R}^6$, $\lambda_2 \in \mathbb{R}^m$ (m is the number of rows in matrix A_2), and $\mu \in \mathbb{R}^6$ are dual variables to the equation constraint (7.23b), the equation constraint (7.23c), and the inequality constraint (7.23d), respectively. The expression (7.24e) can be simplified to

$$\mu = (\mu + B\tau - b)^+, \tag{7.25}$$

where the nonlinear mapping $(\cdot)^+$ is a function which maps negative values to zero and non-negative values to themselves. From Equation (7.24a), $\dot{\pi}$ can be solved as

$$\dot{\pi} = \Lambda_1^{-1}\left(A_1^T\lambda_1 + A_2^T\lambda_2\right). \tag{7.26}$$

Substituting $\dot{\pi}$ in (7.26) to Equations (7.24c) and (7.24d) yields

$$\tau = A_1\Lambda_1^{-1}\left(A_1^T\lambda_1 + A_2^T\lambda_2\right), \tag{7.27}$$

$$\alpha = A_2\Lambda_1^{-1}\left(A_1^T\lambda_1 + A_2^T\lambda_2\right). \tag{7.28}$$

To eliminate τ, we first represent it in terms of λ_1 and μ according to (7.24b) as

$$\tau = -\Lambda_2^{-1}(\lambda_1 + B^T\mu), \tag{7.29}$$

then substitute (7.29) into Equations (7.24c) and (7.27), which results in

$$-\Lambda_2^{-1}(\lambda_1 + B^T\mu) = A_1\Lambda_1^{-1}(A_1^T\lambda_1 + A_2^T\lambda_2), \tag{7.30}$$

$$\mu = (\mu - B\Lambda_2^{-1}\lambda_1 - B\Lambda_2^{-1}B^T\mu - b)^+. \tag{7.31}$$

We use the following dynamics for the solutions of λ_1, λ_2, and μ in Equations (7.28), (7.30), and (7.31):

$$\begin{aligned}
\varepsilon\dot{\lambda}_1 &= -\Lambda_2^{-1}(\lambda_1 + B^T\mu) - A_1\Lambda_1^{-1}(A_1^T\lambda_1 + A_2^T\lambda_2), \\
\varepsilon\dot{\lambda}_2 &= -A_2\Lambda^{-1}A_1^T\lambda_1 - A_2\Lambda^{-1}A_2^T\lambda_2 + \alpha, \\
\varepsilon\dot{\mu} &= -\mu + (\mu - B\Lambda_2^{-1}\lambda_1 - B\Lambda_2^{-1}B^T\mu - b)^+,
\end{aligned} \tag{7.32}$$

where $\varepsilon > 0$ is a scaling factor. Overall, the proposed dynamic neural network has λ_1, λ_2, and μ as state variables and τ in (7.27) as the output, which is expressed as follows:

State equations:

$$\begin{aligned}
\varepsilon\dot{\lambda}_1 &= -\Lambda_2^{-1}\lambda_1 - A_1\Lambda_1^{-1}A_1^T\lambda_1 - A_1\Lambda_1^{-1}A_2^T\lambda_2 \\
&\quad - \Lambda_2^{-1}B^T\mu,
\end{aligned} \tag{7.33a}$$

$$\varepsilon\dot{\lambda}_2 = -A_2\Lambda^{-1}A_1^T\lambda_1 - A_2\Lambda^{-1}A_2^T\lambda_2 + \alpha, \tag{7.33b}$$

$$\varepsilon\dot{\mu} = -\mu + (-B\Lambda_2^{-1}\lambda_1 + \mu - B\Lambda_2^{-1}B^T\mu - b)^+. \tag{7.33c}$$

Output equation:

$$\tau = A_1\Lambda_1^{-1}A_1^T\lambda_1 + A_1\Lambda_1^{-1}A_2^T\lambda_2. \tag{7.33d}$$

About the proposed neural network model (7.33) for the kinematic redundancy resolution problem (7.23) of the parallel manipulator, we have the following remark.

Remark 7.1 In comparison with various feedforward neural network models, such as BP (backpropagation) neural network, RBF (radial basis function) neural network, and ELM (extreme learning machine), the neural network investigated in this chapter falls into the category of recurrent neural networks with dynamic feedbacks. Different from the Kennedy–Chua network, which represents constraints as an extra penalty, the neural network model of this chapter considers the problem in its dual space and is able to reach an optimal solution. The general purpose projected neural network only has a dynamic layer, while the neural network employed in this chapter is organized in a two-layer structure, where the first layer is a dynamic layer with self-feedback, and the second layer is a static one. Projected neural networks are widely used to solve kinematic control of redundant serial manipulators. The dynamic neural network to solve kinematic redundancy in robot motion control also has a two-layered architecture. However, in that model, the hidden layer nodes have an all-to-all connection, while the hidden nodes are partitioned into three cliques in the neural model used in this chapter, where neurons in each clique are connected in an all-to-all manner but the connections between different cliques are local (e.g. the clique associated with the state variable μ is not connected with that associated with the state variable λ_2). The property of local connectivity between hidden cliques also distinguishes this model from other ones using dynamic neural networks for manipulator control.

7.6 Theoretical Results

In this section, we present theoretical results on the proposed neural networks for solving the redundancy resolution problem of parallel manipulators.

7.6.1 Optimality

In this section, we show the equilibrium point of the dynamic neural networks (7.33) ensures that the corresponding output τ given by (7.33d) is identical to the optimal solution of the problem (7.23). On this point, we have the following theorem.

Theorem 7.1 Suppose $(\lambda_1^*, \lambda_2^*, \mu^*)$ is the equilibrium point of the dynamic neural network (7.33). Then the corresponding output τ^* obtained from the output equation (7.33d) is optimal to the constrained programming problem (7.23).

Proof: Letting the right-hand sides of the state equations (7.33a), (7.33b), and (7.33c) be equal to zero, we find the following conditions about the equilibrium point $(\lambda_1^*, \lambda_2^*, \mu^*)$:

$$-\Lambda_2^{-1}\lambda_1^* - A_1\Lambda_1^{-1}A_1^T\lambda_1^* - A_1\Lambda_1^{-1}A_2^T\lambda_2^* - \Lambda_2^{-1}B^T\mu^* = 0, \tag{7.34}$$

$$-A_2\Lambda^{-1}A_1^T\lambda_1^* - A_2\Lambda^{-1}A_2^T\lambda_2^* + \alpha = 0, \tag{7.35}$$

$$-\mu^* + (-B\Lambda_2^{-1}\lambda_1^* + \mu^* - B\Lambda_2^{-1}B^T\mu^* - b)^+ = 0, \tag{7.36}$$

and the corresponding output is:

$$\tau^* = A_1\Lambda_1^{-1}A_1^T\lambda_1^* + A_1\Lambda_1^{-1}A_2^T\lambda_2^*. \tag{7.37}$$

Define an auxiliary value

$$\dot{\pi}^* = \Lambda^{-1}A_1^T\lambda_1^* + \Lambda^{-1}A_2^T\lambda_2^*. \tag{7.38}$$

To show that τ^* is optimal to (7.23), we only need to show that $(\lambda_1^*, \lambda_2^*, \mu^*, \tau^*, \dot{\pi}^*)$ satisfies the KKT condition (7.24) of the optimization problem (7.23), or equivalently the equation set composed of (7.26), (7.27), (7.28), (7.30), and (7.31) (according to the analysis in Section 7.5.1, we can conclude that the equation set composed of (7.26), (7.27), (7.28), (7.30), and (7.31) is equivalent to the KKT condition (7.24)). Comparing the equation set composed of (7.34), (7.35), (7.36), (7.37), and (7.38), and the one composed of (7.26), (7.27), (7.28), (7.30), and (7.31), we find that they are identical and therefore are equivalent. The above procedure implies that the solution $(\lambda_1^*, \lambda_2^*, \mu^*, \tau^*, \dot{\pi}^*)$ is optimal to (7.23). Therefore, we conclude that τ^* is optimal to the problem (7.23). ∎

7.6.2 Stability

In this section, we present theoretical results on the stability of the proposed dynamic neural network model. In Section 7.6.1, we have concluded that the equilibrium point of the neural network (7.33) is optimal solution of (7.23). Generally speaking, a dynamic system may not converge to its equilibrium points. It may happen that a dynamic system evolves towards divergence, oscillation, or even chaos. It is necessary for the proposed neural network to converge for effective computation purposes. Before presenting the convergence results, we first present a lemma about a general projected dynamic system:

$$\dot{u} = -u + P_\Omega(u - F(u)),\qquad(7.39)$$

where $\mu \in \mathbb{R}^l$, Ω is a closed convex set of $\mathbb{R}^l$, and $P_\Omega(\cdot)$ is the projection operator onto the set Ω.

Lemma 7.1 If $\nabla F(u)$ is symmetric and positive semidefinite in $\mathbb{R}^l$, then the dynamic system (7.39) is stable in the sense of Lyapnov and is globally convergent to its equilibrium.

This lemma gives general convergence results on dynamic systems with the presence of projection operators. In our system, the operator $(\cdot)^+$ is also a projection operator, which projects input values to non-negative ones. With Lemma 7.1, it is provable for the following stability results on the proposed model (7.33),

Theorem 7.2 The dynamic neural network (7.33) is stable in the sense of Lyapunov and is globally convergent to the optimal solution of (7.23).

Proof: As we have proved that the output τ associated with the equilibrium points is optimal to problem (7.23) in Theorem 7.1, to draw the conclusion in this theorem we only need to show the convergence of (7.33) to its equilibrium points. To leverage the results presented in Lemma 7.1, we first convert (7.33) into a similar form as (7.39). Define a vector function $F = [F_1^T, F_2^T, F_3^T]^T$, with F_1, F_2, and F_3 defined as follows:

$$F_1 = \Lambda_2^{-1}\lambda_1 + A_1\Lambda_1^{-1}A_1^T\lambda_1 + A_1\Lambda_1^{-1}A_2^T\lambda_2 + \Lambda_2^{-1}B^T\mu,$$
$$F_2 = A_2\Lambda_1^{-1}A_1^T\lambda_1 + A_2\Lambda^{-1}A_2^T\lambda_2 - \alpha,$$
$$F_3 = B\Lambda_2^{-1}\lambda_1 + B\Lambda_2^{-1}B^T\mu + b,\qquad(7.40)$$

and define a set Ω as

$$\Omega = \{(\lambda_1, \lambda_2, \mu), \lambda_1 \in \mathbb{R}^6, \lambda_2 \in \mathbb{R}^m, \mu \in \mathbb{R}^k, \mu \geq 0\},\qquad(7.41)$$

where $\mu \geq 0$ is defined in the piece-wise sense. We also define a new variable

$$x = \begin{bmatrix} \lambda_1^T & \lambda_2^T & \mu^T \end{bmatrix}^T. \tag{7.42}$$

With the above definitions of F, Ω, and x, the proposed neural network (7.33) can be converted as

$$\varepsilon\dot{x} = -x + P_\Omega(x - F(x)). \tag{7.43}$$

Define a new time scale $\phi = \frac{t}{\varepsilon}$. With the new time scale, $\varepsilon\dot{x} = \varepsilon\frac{dx}{dt} = \frac{dx}{d\phi}$, and the neural dynamic (7.43) converts to:

$$\frac{dx}{d\phi} = -x + P_\Omega(x - F(x)), \tag{7.44}$$

which are in the nominal form of the projected dynamic systems. To prove the convergence, we show the symmetricity of ∇F in (7.40) and its positive semidefiniteness:

$$\nabla F = \begin{bmatrix} \Lambda_2^{-1} + A_1\Lambda_1^{-1}A_1^T & A_1\Lambda_1^{-1}A_2^T & \Lambda_2^{-1}B^T \\ A_2\Lambda_1^{-1}A_1^T & A_2\Lambda_1^{-1}A_2^T & 0 \\ B\Lambda_2^{-1} & 0 & B\Lambda_2^{-1}B^T \end{bmatrix}. \tag{7.45}$$

Clearly, ∇F is symmetric. As to the positive semidefiniteness, we decompose ∇F in (7.45) into the following form:

$$\nabla F = \begin{bmatrix} A_1\Lambda_1^{-1}A_1^T & A_1\Lambda_1^{-1}A_2^T & 0 \\ A_2\Lambda_1^{-1}A_1^T & A_2\Lambda_1^{-1}A_2^T & 0 \\ 0 & 0 & 0 \end{bmatrix} + \begin{bmatrix} \Lambda_2^{-1} & 0 & \Lambda_2^{-1}B^T \\ 0 & 0 & 0 \\ B\Lambda_2^{-1} & 0 & B\Lambda_2^{-1}B^T \end{bmatrix}$$

$$= \begin{bmatrix} A_1\Lambda_1^{-\frac{1}{2}} \\ A_2\Lambda_1^{-\frac{1}{2}} \\ 0 \end{bmatrix} \begin{bmatrix} A_1\Lambda_1^{-\frac{1}{2}} \\ A_2\Lambda_1^{-\frac{1}{2}} \\ 0 \end{bmatrix}^T + \begin{bmatrix} \Lambda_2^{-\frac{1}{2}} \\ 0 \\ B_2\Lambda_2^{-\frac{1}{2}} \end{bmatrix} \begin{bmatrix} \Lambda_2^{-\frac{1}{2}} \\ 0 \\ B_2\Lambda_2^{-\frac{1}{2}} \end{bmatrix}. \tag{7.46}$$

The above expression implies that ∇F is indeed positive semidefinite. In summary, as ∇F defined in this proof is symmetric and positive semidefinite, the dynamic system (7.44), i.e. the proposed neural network (7.33), is stable and is convergent to the optimal solution of (7.24) according to Lemma 7.1. ∎

7.6.3 Comparison with Other Control Schemes

In this section, we compare the proposed method with some existing control schemes for a parallel robotic platform.

In [93], feedforward neural network based approaches, including multilayer perception and RBF neural networks, were employed to learn the nonlinear mapping of the robot in its forward kinematics. Thanks to the nonlinear approximation power of feedforward neural networks, the static mapping error can be minimized by feeding enough training data to the model. However, to determine the values on the weights of neural networks, an off-line training procedure is necessary before the method is used. Additionally, due to the lack of consideration for speed limits of each actuator, such constraint may not be satisfied in real applications. Moreover, due to the existence of remaining training error, the control precision largely depends on the size of the training set, and the similarity between the training data and the real ones. In [94], the authors proposed to use self organization mapping to approximate the nonlinearity from joint space to workspace. Due to the required training procedure, and the strong dependence of the performance on training data, this method also inevitably bears similar limitations as [93], such as no compliance to speed limitations and nonoptimality in kinematics. Differently, [95] considered the kinematic model in the design and used the sliding mode method for the tracking control of the Stewart platform. In comparison with methods presented in [93] and [95], this method is inherently an online control scheme, does not require any pre-training, and enjoys a high control precision. However, without considering speed constraints, it still lacks compliance to such a physical restriction in speed. In [96], a fuzzy controller was designed for the online intelligent control of Stewart platforms. This controller is able to reach high-precision tracking of signals in the workspace, but may suffer from similar limitations as the sliding mode based approach [95]. In contrast, since the method presented in this chapter directly considers the problem as an optimization problem, it reaches the optimal solution accordingly. Further, since the speed constraints are taken into account as inequality constraints in the problem formulation, compliance to constraints are thus achieved. The comparisons of different methods for kinematic control of Stewart platforms are summarized in Table 7.1, from which the advantage of the dynamic neural network based approach for the control of Stewart platforms is clear.

7.7 Numerical Investigation

To validate the effectiveness of the proposed approach, in this section we apply the neural network model to the redundancy resolution of a physically constrained Stewart platform.

7.7.1 Simulation Setups

In the simulation, we consider a Stewart platform with the leg connectors on the mobile platform locating around a circle with radius 1.0 m at $b'_1 = [0.7386, 0.1302, 0]$, $b'_2 = [0.7386, -0.1302, 0]$, $b'_3 = [-0.4821, 0.5745, 0]$, $b'_4 = [-0.2565, 0.7048, 0]$,

Table 7.1 Comparisons of different methods for kinematic control of Stewart platforms.

	Optimality	Online vs. Off-line	Pre-training	Compliance to extra constraints	Precision
Approach in [93]	No	Off-line	Yes	No	Low
Approach in [94]	No	Off-line	Yes	No	Low
Approach in [95]	No	Online	No	No	High
Approach in [96]	No	Online	No	No	High
This chapter	Yes	Online	No	Yes	High

$b_5' = [-0.2565, -0.7048, 0]$, and $b_6' = [-0.4821, -0.5745, 0]$ in the platform coordinate, and the leg connectors locating around a circle with radius 0.75 m at $a_1 = [0.3750, 0.6495, 0]$, $a_2 = [0.3750, -0.6495, 0]$, $a_3 = [-0.7500, 0.0000, 0]$, $a_4 = [0.3750, 0.6495, 0]$, $a_5 = [0.3750, -0.6495, 0]$, and $a_6 = [-0.7500, 0.0000, 0]$ on the fixed base. For simplicity, the end-effector is put at the origin of the platform coordinate (this can always be achieved by defining the platform coordinate with its origin at the end-effector position). In the situation for position tracking in three-dimensional space, the total redundancy is 3 as the input dimension is 6 while the output dimension is 3.

In the simulation, we use the tracking error, defined as the difference between the desired position and the real position at time t, to measure the tracking performance in the circular motion tracking, infinity-sign curve tracking, and the square motion tracking situation.

(a)

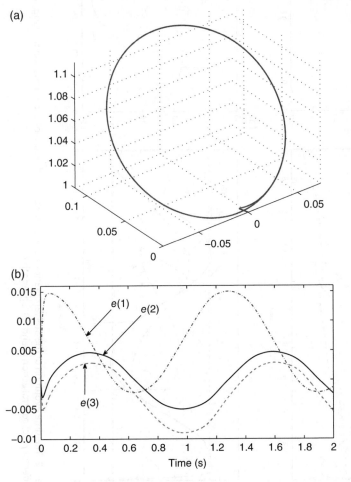

(b)

Figure 7.1 Tracking of a circular motion. (a) The trajectory of the end-effector and (b) the time history of the position tracking error.

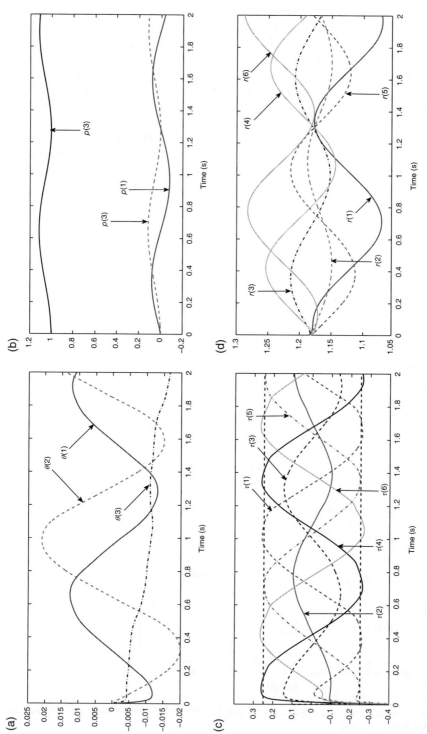

Figure 7.2 The time evolution of the Stewart platform state variables in the case of circular motion tracking. (a) Orientation of the platform θ; (b) position of the end-effector p; (c) control action τ; and (d) leg length r.

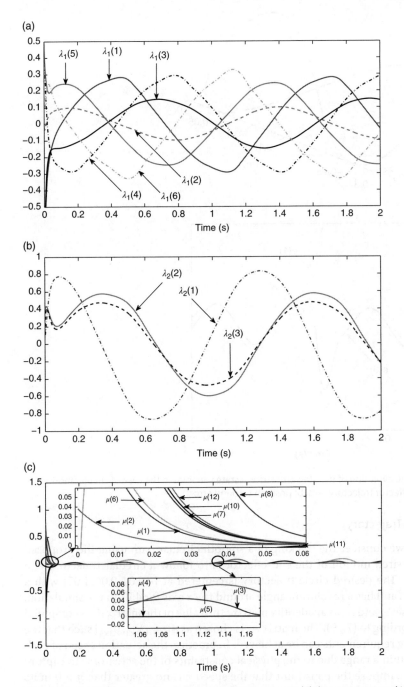

Figure 7.3 Simulation results on motion trajectories, manipulability measures and joint-angle profiles of PUMA 560 synthesized by the scheme presented in [71] with its end-effector fixed at [0.55, 0, 1.3] m in the workspace. (a) State variable λ_1; (b) state variable λ_2; and (c) state variable μ.

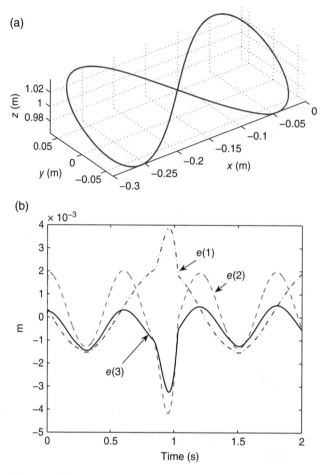

Figure 7.4 The time evolution of the Stewart platform state variables in the case of circular motion tracking. (a) End-effector trajectory and (b) position tracking error.

7.7.2 Circular Trajectory

In this section, we consider the tracking of a smooth circular path using the proposed method. The desired motion of the end-effector is to follow a circular trajectory at a speed of 2 m/s. The desired circle trajectory is centered at $[-0.04, 0.06, 1.05]$ with a radius of 0.08 m, and has a revolution angle around the x axis for $45°$. In the simulation, we simply choose Λ_1 and Λ_2 as an identity matrix. The value of the matrix A_1 is computed in real time according to (7.19). The matrix A_2 is chosen as $A_2 = [I_{3\times3}, 0_{3\times3}]$ such that the position tracking requirements can be achieved. In practice, the actuation speed of each leg is limited within a range due to the physical constraints of the actuators. To capture this property, we impose the constraint that the speed τ is no greater than $\eta > 0$ in its absolute value, i.e. $|\tau_i| \le \eta$ for $i = 1, 2, ..., 6$, which is equivalent to $-\eta \le \tau_i \le \eta$. Organizing to a matrix form yields an inequality in the form of (7.23d) with $B = [I_6 \times 6, I_6 \times 6]^T$, $b = \eta\mathbf{1}_{12}$ with $\mathbf{1}_{12}$ representing a twelve-dimensional vector with all entries equal one. In the simulation, we set the speed bound $\eta = 0.25$ m/s. The scaling factor ε is chosen

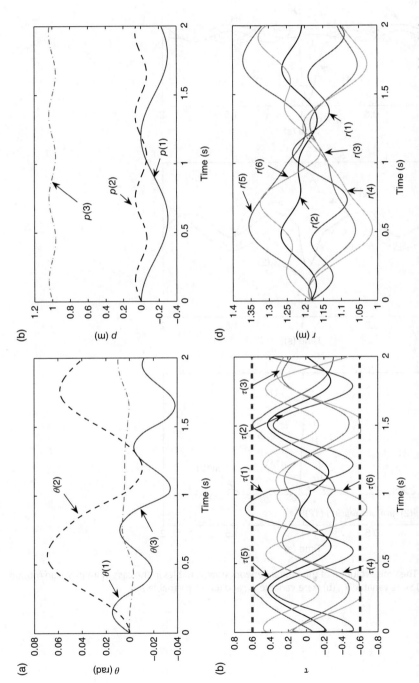

Figure 7.5 The time evolution of the Stewart platform state variables in the case of infinity-sign motion tracking. (a) Orientation of the platform θ; (b) position of the end-effector p; (c) control action τ; and (d) leg length r.

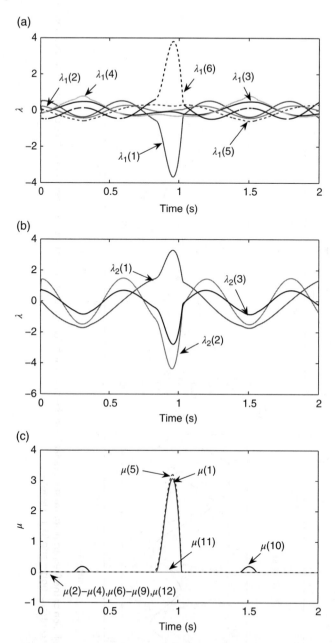

Figure 7.6 The time evolution of the neural network state variables in the case of infinity-sign motion tracking. (a) State variable λ_1; (b) state variable λ_2; and (c) state variable μ.

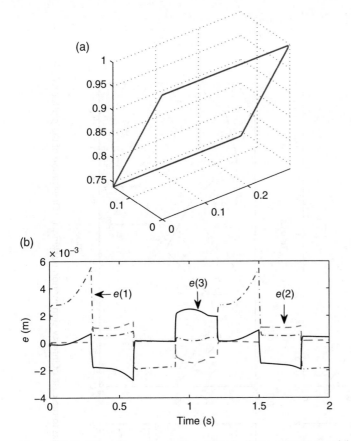

Figure 7.7 Tracking of a square motion. (a) End-effector trajectory and (b) position tracking error.

as $\varepsilon = 0.01$. The tracking results are shown in Figure 7.1 by running the simulation for 2 s. As shown in the Figure 7.1a, the end-effector successfully tracks the circular path with a small tracking error (as shown in Figure 7.1b, where $e(1)$, $e(2)$, and $e(3)$ denote the components of the position tracking error e, respectively, along the x, y and z axes of the base frame; the errors are less than 0.015 m in amplitude). The circular-path following experiments demonstrated the capability of the proposed dual neural network for online resolving of kinematic redundancy of physically constrained manipulators.

Considering the input motion for the six legs, the Figure 7.2 shows the time profile of the Stewart platform state variables, i.e. the three Euler orientations of the mobile platform (Figure 7.2a), the end-effector position (Figure 7.2b), the speed of each leg (Figure 7.2c) and the length of each leg (Figure 7.2d). The attached moving frame to the upper platform is exactly in the middle and the $p(1)$ and $p(2)$ coordinates started from zero and the $p(3)$ oscillates between almost 0.005 m and −0.006 m. The $p(2)$ coordinate started from approximately 0.125 m which is the altitude of the upper platform before the motion of the actuators. It was observed that the harmonic response is repeated every 1.25 s or 2π after that the cycle is repeated again. From Figure 7.2a, it is observed that there is a small drift from the neutral position. Despite the presence of this drifting in orientation, the desired motion, which is given in terms of end-effector positions,

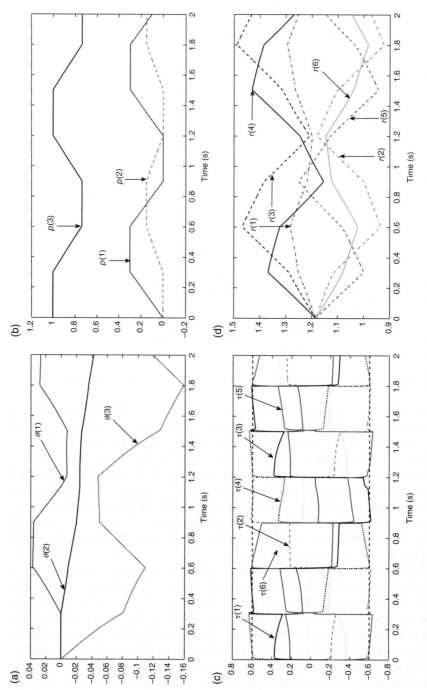

Figure 7.8 The time evolution of the Stewart platform state variables in the case of infinity-sign motion tracking. (a) Orientation of the platform θ; (b) position of the end-effector p; (c) control action τ; and (d) leg length r.

is still achieved, as shown in Figure 7.3. This in turn validates the robustness of the proposed scheme in the presence of orientation drifting. The straight dashed line in Figure 7.2(c) depicts the bound ± 0.25 ($\eta = 0.25$), which is the bound for the action speed. It can be observed that τ converges to the region $[-0.25, 0.25]$ very fast and stays approximately inside this region through the runtime, except for some short period (e.g. $\tau(6)$ at around time $t = 1.7$ s) due to the dynamics of the desired motion. The neural network state variables are plotted in Figure 7.3. From this figure, we can clearly observer the dynamic evolution of the neural activities. It is noteworthy that the neural activities do not converge to a constant value. Instead, they vary with time as they are utilized to compensate and regulate the dynamic motion of the robot.

7.7.3 Infinity-Sign Trajectory

In this section, we consider the tracking of an infinity-sign curve using the presented method. Specifically, we consider the tracking at a constant speed of 1 m/s. The desired trajectory is obtained by rotating a planar infinity-sign curve analytically expressed as $x^4 = a^2(x^2 - y^2)$ with $a = 0.1$ 5m around the x axis for 45°. The center of this curve locates at $[-0.04, 0.06, 1.05]$. The parameters, Λ_1, Λ_2, A_1, A_2, η, B, b, and ε are chosen with the same value as in the circular trajectory tracking situation. Simulation results are obtained by running it for 2 s. As shown in the Figure 7.4, the end-effector successfully tracks the desired path with a small tracking error. Figure 7.5 show the time profile of the Stewart platform state variables, including the three Euler orientations of the mobile platform (Figure 7.5a), the end-effector position (Figure 7.5b), the speed of each leg (Figure 7.5c), and the length of each leg (Figure 7.5d). The neural network state variables are plotted in Figure 7.6. Due to the dynamic changing of the neural states, the nonlinearity in the Stewart platform is successfully compensated and an accurate tracking performance is thus achieved.

7.7.4 Square Trajectory

In this section, we investigate the square trajectory tracking using the proposed approach. Different from the case of smooth circular motion tracking, the desired square path is non-smooth at the four corners and poses challenges to the controller on its real-time performance. In the simulation, the desired motion of the end-effector is to follow a square trajectory, which is centered at $[0.15, 0.075, 0.74]$ with an edge length of 0.08 m, at the desired speed of 1.0 m/s. The square has a revolution angle around the x axis for 60°. We choose the parameters Λ_1 and Λ_2, A_2 and B to have the same values as in the situation for circular motion for simplicity. The parameter A_1 is computed in real time according to (7.19). The speed limit bound η is chosen as $\eta = 0.6$ m/s, and b is accordingly chosen as $b = 0.61_{12}$. The scaling factor ε is chosen as $\varepsilon = 0.001$. The tracking results are shown in Figure 7.7 by running the simulation for 2 s. As shown in the Figure 7.7a, the end-effector successfully tracks the square motion with a small tracking error (as shown in Figure 7.7b, where $e(1)$, $e(2)$, and $e(3)$ denote the components of the position tracking error e, respectively, along the x, y and z axes of the base frame; the errors are less than 0.006 m in amplitude). Note that the error curves have jerks at time $t = 0.3, 0.6, 0.9, 1.2, 1.5,$ and 1.8 s, which is a result of the velocity switching from one direction to another one around the corner of the

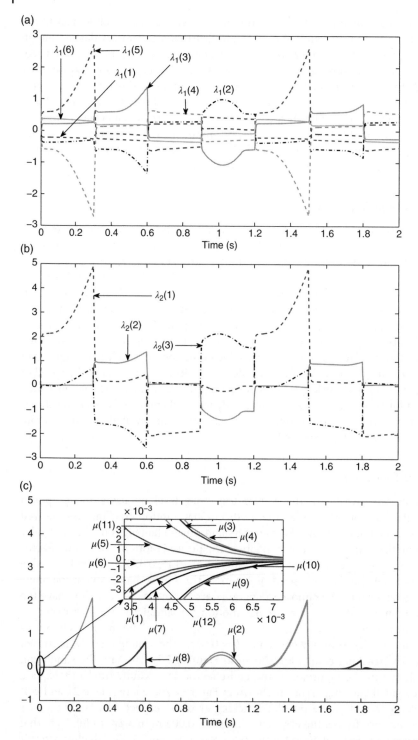

Figure 7.9 The time evolution of the neural network state variables in the case of square motion tracking. (a) State variable λ_1; (b) state variable λ_2; and (c) state variable μ.

square. Despite the existence of the jerks, due to the nonlinear feedback mechanism in the neural network, the errors reduce swiftly to a very low value (much lower than 0.001 m as shown in Figure 7.7b after the jerk). The above observation in turn validates the effectiveness of the proposed neural scheme in dealing with non-smooth tracking problems.

The real-time values on the Stewart platform states are shown in Figure 7.8. Similar to the situation for circle tracking, the presence of drifting in orientations, as observed in Figure 7.8a, does not affect the tracking performance. The end-effector position evolves without drifting (Figure 7.8b), and remains a very small error from its desired trajectory (Figure 7.8b). Figure 7.8c shows the time profile of the control action, which is the speed of each leg. It is clear that the control action is approximately regulated inside the range $[-0.6, 0.6]$, which in turn validates the effect of the proposed solution in fulfilling the inequality regulation. The dynamic evolution of the neural activities is shown in Figure 7.9.

7.8 Summary

In this chapter, a dynamic neural network is designed to solve the redundancy resolution problem of Stewart platforms. The physical constraints and optimization criteria are rigorously modeled as a constrained quadratic programming problem. To solve this problem in real time, a recurrent neural network is proposed to reach the equality constraints, inequality constraints, and optimality criteria simultaneously. Rigorous theoretical proofs are supplied to verify the convergence of the proposed model. Simulation results validate the effectiveness of the proposed solution.

8

Neural Network Based Learning and Control Co-Design for Stewart Platform Control

8.1 Introduction

Kinematically redundant manipulators [39] are those manipulators that prove to have sufficiently higher degrees of freedom (DOFs) than required for positioning and for orientation of the platform. With the advances in the field of robotic technologies, robotic manipulators are widely used in the applications of factory automation which are required to carry out continuous and delayed work, such as lifting and transporting radioactive substances and executing the work in hazardous, scattered or packed environments. In comparison with nonredundant manipulators, redundant ones offer extra DOFs [75], and are often used to improve the dexterity, in order to work efficiently by avoiding collisions with obstacles. Research in the field of kinematically redundant manipulators has gained popularity due to their ability to avoid obstacles, their internal singular configurations, and their ability to optimize the performance of the workspace and the end-effector motion task. Among various types of redundant manipulators, parallel ones, which usually feature higher rigidity, higher precision and higher response speed than serial ones, have received popular applications in flight simulators, electrostatic magnetic lenses, etc. However, how to efficiently control the motion of redundant parallel manipulators, especially in the situation with parameter uncertainties, or even unknown, is of great practical significance and remains a challenging research problem. Parallel redundant manipulators are broadly classified as parallel manipulators where the task space coordinates are lower than the number of actuators. These manipulators are found in many industrial applications such as robotics arms, surgical robots, and so on. These manipulators offer greater advantages when incorporated because they make the structure flexible, faster, and lighter thereby improving the Cartesian stiffness and optimizing the distribution of the force. Due to the advantages of high speed and high acceleration, parallel manipulators have been studied and implemented widely.

Redundant manipulators follow similar dynamic equations as for the serial ones. Therefore, extending design studies of serial manipulators to parallel manipulators is fairly obvious. However, designing the dynamics of the parallel manipulator is more complicated. Moreover, it is quite challenging to identify the unknown parameter which describes the dynamic behavior or properties of the system. Hence, the control

Kinematic Control of Redundant Robot Arms Using Neural Networks, First Edition.
Shuai Li, Long Jin and Mohammed Aquil Mirza.
© 2019 John Wiley & Sons Ltd. Published 2019 by John Wiley & Sons Ltd.

schemes of the existing traditional methods for serial manipulators cannot be extended in real-time control applications to parallel redundant manipulators.

Parallel manipulators are confined to age-old and basic problems of identification and classification of singularities. A lot of work is developed using mathematical tools borrowed from serial manipulators for local analysis and to resolve the problem of singularities. Gosselin was the first to define, study, analyze and report the singularity problem for the closed-loop kinematic chains [97]. The structure and behavior of the singularity problem for the parallel manipulator is indeed complex and challenging. Many studies are incorporated to address the kinematic manipulability measure for design and control of parallel mechanisms. Recurrent neural networks, as a powerful parallel computation method, are proven effective and efficient for the applications of real-time solutions to the inverse kinematics problem. In the literature of the past decade, a variety of dynamical system solvers have been proposed to resolve the problems of online constrained quadratic programming, including the primal dual network, Lagrange neural networks, the gradient network, and the projected network. There are also traditional approaches which consider joint and velocity constraints. For expressing a general solution in the form of redundant join velocities, the Gram–Schmidt orthogonalization procedure is utilized.

Constraints in soft computing techniques introduce two main categories of difficulties in obtaining the solution to the problem. First, the challenge is of the independence where the coordinates are not independent; and secondly, a priori information of the constraints forces is not sufficiently provided and they are regarded among the unknowns of the system. Hence, control of the Stewart platform as a constraint system becomes complicated due to the complex nature of the neural dynamics. Another limiting factor in conventional robotic manipulation research using dual neural network approaches is the requirement for a design model, which involves constructing a mathematical model that highlights the controlled dynamics of the model. The initial stage of the design usually requires the interd dependence between different parts and their historical dependence on previous states to be established. A later stage involves the design of the analytical controller with the mathematically modeled system dynamics. Although the designed control law gives a promising performance for the mathematical model, this might not be the case in real-time applications as the exact representation of the model is hard to obtain. This may be due to various reasons, e.g. the sheer complexity of the designed model or the uncertainty involved in the area. However, modeling of the feedback control for the physical system brings about the tradeoff between the ease of modeling and the precision of the physical system in matching. Due to the improvement of the parameters in the controller depending upon the convergence and stability factors, adaptive techniques usually demonstrate outstanding results in the face of complex systems. This motivates us to devise an adaptive and model-free neural controller to steer the motion of a Stewart platform.

In this chapter we aim to close the gap between the two research areas of model-free recurrent neural networks for machine learning and model-based dual neural networks for accurate model control. The proposed network interacts with the learning and control parts. An excitation noise is added to avoid the learning degradation. This deliberate design offers precise convergence of the estimated variables to their true values. The stability of the network is proved theoretically and by simulations and it is proved that the bounded error is found to be arbitrarily small by scaling the additive noise.

8.2 Kinematic Modeling of Stewart Platforms

A Stewart platform, as sketched in Figure 8.1, consists of a 6-DOF platform comprising two plates, namely a fixed base plate and a flexible or moving top plate, which is in turn connected to a series of prismatic actuators and passive joints. Each of the prismatic actuators is connected by a spherical joint to the base plate. A base plate is connected by universal joints to each of its actuators. This specific arrangement of actuators and joints allows the top moving plate to move on either side depending upon the lengths of the prismatic actuators or leg joints.

8.2.1 Geometric Relation

There are two coordinate systems associated with a Stewart platform, namely a base coordinate system which is fixed as a global system and a moving coordinate system of the platform. We use x' to distinguish a variable defined in the base coordinate system from the corresponding one defined as x in the global coordinate. In Figure 8.2, the position vectors b_i indicate the position of the center of the universal joint of the leg. Thus, a_i, defined as the position vectors, represent the moving platform positions in global coordinates of the base's ith connection point. The vector $d_i = b_i - a_i$ represents the ith leg of the actuator pointing from the base to the platform. The global coordinates and the platform coordinates are fixed to the base and the mobile platform, respectively. $a_i \in \mathbb{R}^3$ for $i = 1, 2, \cdots, 6$ dictates the position in global coordinates of the ith connection point on the base. $b_i' \in \mathbb{R}^3$ for $i = 1, 2, \cdots, 6$ denotes the position in platform coordinates of the ith connection point on the platform. Hence b_i are defined to represent the global coordinate (Figure 8.2). $d_i = b_i - a_i$ for $i = 1, 2, \cdots, 6$ is defined to represent the ith leg vector from the base to the platform. For $x' \in \mathbb{R}^3$ in platform coordinates, the corresponding global coordinates $x \in \mathbb{R}^3$ can be derived after a translational and rotational transformation:

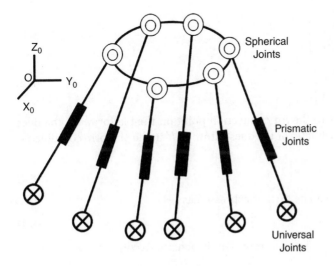

Figure 8.1 Schematic of a Stewart platform.

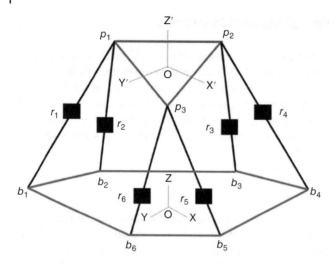

Figure 8.2 Stewart platform geometric representation. The gray triangle at the top and the gray hexagon at the bottom represent the moving top plate and the fixed base plate, respectively.

$$x = p + Qx',$$

(8.1)

where $p = [x_p, y_p, z_p]^T \in \mathbb{R}^3$ represents the global coordinates of the origin position in the platform coordinate, and corresponding to the translational transformation, $Q \in \mathbb{R}^3$ is the rotational matrix, which is predominantly defined by the Euler angles $\theta = [\theta_x, \theta_y, \theta_z]^T \in \mathbb{R}^3$,

$$Q = Q_z Q_y Q_x,$$

$$Q_x = \begin{bmatrix} 1 & 0 & 0 \\ 1 & \cos\theta_x & \sin\theta_x \\ 1 & -\sin\theta_x & \cos\theta_x \end{bmatrix}$$

$$Q_y = \begin{bmatrix} \cos\theta_y & 0 & -\sin\theta_y \\ 0 & 1 & 0 \\ \sin\theta_y & 0 & \cos\theta_y \end{bmatrix}$$

$$Q_z = \begin{bmatrix} \cos\theta_z & \sin\theta_z & 0 \\ -\sin\theta_y & \cos\theta_y & 0 \\ 0 & 0 & 1 \end{bmatrix}.$$

(8.2)

Following Equation (8.1), as to the ith connection point on the platform, i.e. the ones with $x = b_i$ in the global coordinates or the ones with $x' = b_i'$ in the platform coordinates, we have

$$b_i = p + Qb_i',$$

(8.3)

Therefore, the ith leg vector can be further expressed as

$$d_i = b_i - a_i = p + Qb_i' - a_i.$$

(8.4)

For the vector d_i, we define $r_i = \| d_i \|$ to represent its length. Accordingly, we have

$$r_i = \| p + Qb_i' - a_i \|.$$

(8.5)

It is to be observed that both a_i and b_i are constants. Henceforth, they can be derived by the geometric structure. $p = [x_p, y_p, z_p]^T$ denotes the coordinate frame which defines the translation of the platform, and Q represents a function of the Euler angles $\theta = [\theta_x, \theta_y, \theta_z]^T$, denoting the rotation of the Stewart platform. The right-hand side of (8.5) depends on the pose variables of the Stewart platform $\pi = [x_p, y_p, z_p, \theta_x, \theta_y, \theta_z]^T \in \mathbb{R}^6$ while the left-hand side is the length of the leg, which is in turn controlled for actuation. Hence we highlight that in this way, Equation (8.5) for $i = 1, 2, \cdots, 6$ defines the relationship of kinematics between the actuation variables and the pose variables. Now, for a defined six-dimensional reference position, the desired leg length r_i can be directly obtained from (8.5). However, in real-time applications, the reference positions are usually not defined as six-dimensional. This could be explained by an example from the medical industry: in surgical applications which include a Stewart platform, doctors only care about the position of an end-effector on the platform, and are not concerned about its orientation. Owing to this fact, usually in these scenarios the reference is three-dimensional and therefore we have three additional DOFs as redundancy. For these kind of challenges, we ultimately have an infinite number of feasible solutions of r_i for $i = 1, 2, \cdots, 6$ to reach the reference. Among all of the feasible options and solutions, we may be able to identify a unique solution, which outperforms others and existing solutions in terms of certain criteria of optimization. This intuitive and mathematical approach analyzes and motivates the modeling of the optimization problem and identifies the optimal solution for the improvised performance. Due to presence of the nonlinearity of Equation (8.5), treating Equation (8.5) directly is technically impractical and prohibitive. Hence, we formulate the problem in terms of velocity space to explore the linearity approximation rather than studying the problem in position space for a direct solution.

8.2.2 Velocity Space Resolution

For simplicity, Equation (8.5) is rearranged as follows:

$$r_i^2 = (p + Qb_i' - a_i)^T (p + Qb_i' - a_i). \tag{8.6}$$

We define, represent and obtain the velocity space relations by computing the time derivative on both sides of (8.6), which yields

$$r_i \dot{r}_i = (p + Qb_i' - a_i)^T (\dot{p} + \dot{Q}b_i' + Q\dot{b}_i' - \dot{a}_i) = (p + Qb_i' - a_i)^T (\dot{p} + \dot{Q}b_i'). \tag{8.7}$$

Recall that both a_i and b_i' are constants and their time derivatives, $\dot{a}_i$ and $\dot{b}_i'$, are equivalent to zero. For the rotational matrix Q, we consider the the above discussed preliminary equations, and the following represents the time derivative property

$$\dot{Q}Q^T = \begin{bmatrix} 0 & -\dot{\theta}_z & \dot{\theta}_y \\ \dot{\theta}_z & 0 & -\dot{\theta}_x \\ -\dot{\theta}_y & \dot{\theta}_x & 0 \end{bmatrix} = \begin{bmatrix} \dot{\theta}_x \\ \dot{\theta}_y \\ \dot{\theta}_y \end{bmatrix}_\times = \dot{\theta}_\times. \tag{8.8}$$

In this way, the moving platform rotation matrix coordinate system with respect to base platform is achieved. However, the position vector specified at the origin of the moving platform denotes the translation vector with respect to the base platform.

$$\dot{Q} = \dot{\theta}_\times (Q^T)^{-1} = \dot{\theta}_\times Q. \tag{8.9}$$

Substituting (8.9) into (8.7) yields

$$
\begin{aligned}
r_i \dot{r}_i &= (p + Qb'_i - a_i)^{\mathrm{T}}(\dot{p} + \dot{\theta}_\times Qb'_i) \\
&= d_i^{\mathrm{T}}(\dot{p} + \dot{\theta}_\times Qb'_i) \\
&= d_i^{\mathrm{T}}\dot{p} + (\dot{\theta} \times Qb'_i)^{\mathrm{T}} d_i \\
&= d_i^{\mathrm{T}}\dot{p} + ((Qb'_i) \times d_i)^{\mathrm{T}}\dot{\theta} \\
&= [d_i^{\mathrm{T}} \quad ((Qb'_i) \times d_i)^{\mathrm{T}}]
\begin{bmatrix} \dot{p} \\ \dot{\theta} \end{bmatrix}.
\end{aligned}
\tag{8.10}
$$

As mentioned, in the above equation, Equation (8.4) is used for the derivation in the second and the penultimate lines, respectively. Noticing that $r_i = \| d_i \| > 0$ could be guaranteed by the mechanical structure, we have the following result

$$
\dot{r}_i = \frac{1}{r_i}[d_i^{\mathrm{T}} \quad ((Qb'_i) \times d_i)^{\mathrm{T}}]
\begin{bmatrix} \dot{p} \\ \dot{\theta} \end{bmatrix}
= \frac{1}{r_i}[d_i^{\mathrm{T}} \quad ((Qb'_i) \times d_i)^{\mathrm{T}}]\dot{\pi}.
\tag{8.11}
$$

For the six-dimensional vector $r = [r_1, r_2, \cdots, r_6]^{\mathrm{T}}$, we have the compact matrix form as follows

$$
\dot{r} = A_1 \dot{\pi},
\tag{8.12}
$$

where

$$
A_1 =
\begin{bmatrix}
\frac{1}{r_1}d_1^{\mathrm{T}} & \frac{1}{r_1}((Qb'_1) \times d_1)^{\mathrm{T}} \\
\frac{1}{r_2}d_2^{\mathrm{T}} & \frac{1}{r_2}((Qb'_2) \times d_2)^{\mathrm{T}} \\
\cdots & \cdots \\
\frac{1}{r_6}d_6^{\mathrm{T}} & \frac{1}{r_6}((Qb'_6) \times d_6)^{\mathrm{T}}
\end{bmatrix}
=
\begin{bmatrix}
\frac{1}{r_1}(p + Qb'_1 - a_1)^{\mathrm{T}} & \frac{1}{r_1}((Qb'_1) \times (p - a_1))^{\mathrm{T}} \\
\frac{1}{r_2}(p + Qb'_2 - a_2)^{\mathrm{T}} & \frac{1}{r_2}((Qb'_2) \times (p - a_2))^{\mathrm{T}} \\
\cdots & \cdots \\
\frac{1}{r_6}(p + Qb'_6 - a_6)^{\mathrm{T}} & \frac{1}{r_6}((Qb'_6) \times (p - a_6))^{\mathrm{T}}
\end{bmatrix}.
$$
$$
\tag{8.13}
$$

Equation (8.12) projects the kinematic relation of a Stewart platform with 6 DOFS from the velocity of the pose variables to the speed of the legs.

8.3 Recurrent Neural Network Design

8.3.1 Problem Formulation from an Optimization Perspective

In this section, we introduce a numerical and nonlinear gradient decent optimization method to resolve the real kinematic parameters from the measurement data. The digital indicators are defined as measurement devices in order to rectify and verify the location of the end-effector of the Stewart platform. This in turn can determine the error between the desired and actual locations. The equation corresponding to kinematics of a parallel Stewart platform can be expressed as follows:

$$
r_i = \| p + Qb'_i - a_i \|
\tag{8.14}
$$

where a_i represents the ith position vector in the mobile platform with respect to A and b_i represents the ith position vector with respect to the base B. r_i represents the ith link length. This equation denotes the peculiar inverse kinematic equation of a Stewart platform. In general, it is infeasible to derive a forward kinematic model for the parallel manipulators due to nonavailability of a closed form solution, and some numeral algorithms must be incorporated to derive the parameters of the forward kinematics.

The nonlinear or nonconvex relationship between r_i and π to the position variable is affine to $\dot{\pi}$. We observed in our analysis that $\dot{r}$ can be solved directly from the compact matrix form equation (8.12). The reference velocity constrained in reduced dimensions for the equality model is defined as

$$\alpha = A_2 \dot{\pi}, \tag{8.15}$$

where $\alpha \in \mathbb{R}^m$ is the reference vector with $0 < m < 6$. The pre-given transformation matrix is $A_2 \in \mathbb{R}^{m \times 6}$. The optimized solution is defined as follows:

$$\min_{(\dot{\pi}, \tau)} \frac{1}{2} \dot{\pi}^T \Lambda_1 \dot{\pi} + \frac{1}{2} \tau^T \Lambda_2 \tau, \tag{8.16}$$

where the symmetric matrices $\Lambda_1 \in \mathbb{R}^{6 \times 6}$ and $\Lambda_2 \in \mathbb{R}^{6 \times 6}$ are both constant and positive definite, and the speed of the platform legs to be control is denoted as $\tau = \dot{r}$. In practice, the term $\dot{\pi}^T \Lambda_1 \dot{\pi}$, which is in a quadratic form, specifies the kinematic energy when choosing Λ_1 properly, and the input power can be characterized by $\tau^T \Lambda_2 \tau$. The formulation of the objective function is consistent with the convention of control theory in defining quadratic cost functions. The actuator can directly change the value of the decision variable τ. Its value is under physical constraints, which are modeled as inequalities in the following form,

$$B\tau \leq b. \tag{8.17}$$

In this expression, $B \in \mathbb{R}^{k \times 6}$ and $b \in \mathbb{R}^k$ with k as an integer. It is noteworthy that it is not imposed on the variable $\dot{\pi}$ for the constraint as in the planning stage it usually has already been specified. In summary, with Equation (8.16) as the object function, and (8.17) as physical constraints, and also with the nonlinear dynamic equation constraints, a constrained programming can thus be formulated to solve the control:

$$\min_{(\dot{\pi}, \tau)} \quad \frac{1}{2} \dot{\pi}^T \Lambda_1 \dot{\pi} + \frac{1}{2} \tau^T \Lambda_2 \tau, \tag{8.18a}$$

$$\text{s.t.} \quad \tau = A_1 \dot{\pi}, \tag{8.18b}$$

$$\alpha = A_2 \dot{\pi}, \tag{8.18c}$$

$$B\tau \leq b. \tag{8.18d}$$

Since there are two types of constraints present, namely equality and inequality constraints in (8.18), it is not feasible to solve the optimizaton analytically. Incorporating traditional approaches incurs an extra penalty in terms formed by the constraints to the objective function. Resolving the problem numerically using the approach of gradient descent along the new objective function is also expensive. Hence we conclude that penalty-based approaches are expensive and can only reach an approximate solution of the problem and therefore are not feasible to tackle error-sensitive applications. Hence, it is worth attempting to devise a dynamic differential equation for the type of neural network to approach the solution iteratively.

8.3.2 Neural Network Dynamics

In this section, we present the neural network model used in this chapter. This is a dynamic neural model that can be described by an ordinary differential equation. The dynamics is as follows:

State equations:

$$\varepsilon\dot{\lambda}_1 = -\Lambda_2^{-1}\lambda_1 - \hat{A}_1\Lambda_1^{-1}\hat{A}_1^T\lambda_1 - \hat{A}_1\Lambda_1^{-1}A_2^T\lambda_2$$
$$- \Lambda_2^{-1}B^T\mu, \tag{8.19a}$$

$$\varepsilon\dot{\lambda}_2 = -A_2\Lambda^{-1}\hat{A}_1^T\lambda_1 - A_2\Lambda^{-1}A_2^T\lambda_2 + \alpha, \tag{8.19b}$$

$$\varepsilon\dot{\mu} = -\mu + (-B\Lambda_2^{-1}\lambda_1 + \mu - B\Lambda_2^{-1}B^T\mu - b)^+. \tag{8.19c}$$

Output equation:

$$\bar{\tau} = \hat{A}_1\Lambda_1^{-1}\hat{A}_1^T\lambda_1 + \hat{A}_1\Lambda_1^{-1}A_2^T\lambda_2, \tag{8.19d}$$

$$\tau = \bar{\tau} + w, \tag{8.19e}$$

$$\zeta\dot{\hat{A}}_1 = -(\hat{A}_1\dot{\pi} - \tau)\dot{\pi}^T, \tag{8.19f}$$

where $\zeta > 0$ is a scaling factor. By replacing A_1 with $\hat{A}_1$ as obtained in (8.1), the whole system obtained so far can be expressed as

$$\zeta\dot{\hat{A}}_1 = -(\hat{A}_1\dot{\pi} - \tau)\dot{\pi}^T. \tag{8.20}$$

Consider a special case when the initial value of $\hat{A}_1$ and π at time $t=0$ are both set at zero. In this situation, the immediate derivative of state variables can be obtained as $\dot{A}_1 = 0$. Then

$$\zeta\dot{A}_1 = -(A_1\dot{\pi} - \tau)\dot{\pi}^T. \tag{8.21}$$

Subtracting (8.20) from (8.21) yields

$$\zeta(\dot{\hat{A}}_1 - \dot{A}_1) = (A_1 - \hat{A}_1)\dot{\pi}\pi^T. \tag{8.22}$$

Since, $\hat{A}_1 - A_1 = \tilde{A}_1$,

$$\zeta\dot{\hat{A}}_1 = (-\tilde{A}_1)\dot{\pi}\pi^T. \tag{8.23}$$

The state variables of the neural network are shown in Figure 8.3. This figure depicts the dynamic redundancy of the neural network.

8.3.3 Stability

Stability is the most important issue in the dynamic systems. Nonstable systems may oscillate or even diverge. In this section we discuss the stability of the proposed architecture in learning convergence.

Define $\hat{A}_1 - A_1 = \tilde{A}_1$ and use $V_1 = \parallel \tilde{A}_1 \parallel_F^2 = \text{trace}(\tilde{A}^T\hat{A}_1)/2$, where $\parallel . \parallel_F$ denotes the Frobenius norm of a matrix, as an estimation error metric. The time derivative of V_1 along the system dynamic is $\dot{V}_1 = \text{trace}(\tilde{A}^T\dot{\hat{A}}_1) = -\frac{1}{\zeta}\text{trace}[\tilde{A}^T(\hat{A}_1\dot{\pi} - \tau)\dot{\pi}^T]$.

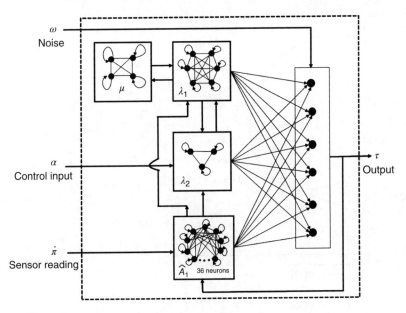

Figure 8.3 Architecture of the proposed neural network.

Note that the equality trace $(E(AB)) = \text{trace}(E(BA))$ for any A and B of appropriate size is utilized in the above derivation. Thus, $\dot{V}_1 = -\frac{1}{\zeta}\text{trace}(\tilde{A}\dot{\pi}^{\mathrm{T}})$. For $C = \tilde{A}_1\dot{\pi}$ $\dot{V} = -\frac{1}{\zeta}\text{trace}(C^+C) = -\frac{1}{\zeta}\|\tilde{A}_1\dot{\pi}\|_F^2 \leq 0$. Note that $V \geq 0$ and is monotonically decreasing according to the above equations.

We have

$$\tau = A_1\dot{\pi},$$
$$\tau = \overline{\tau} + w,$$
$$\overline{\tau} + w = A_1\dot{\pi},$$
$$\dot{\pi} = A^+(\overline{\tau} + w).$$

We know that $\tilde{A}\dot{\pi} = 0$, we can multiply $\tilde{A}_1^{\mathrm{T}}$, $(\overline{\tau} + w)^{\mathrm{T}}$ and computing the expected value yields

$$E[\text{trace}[\tilde{A}_1^{\mathrm{T}}\tilde{A}_1\tilde{A}_1^+(\overline{\tau} + w)(\overline{\tau} + w)^{\mathrm{T}}]] = 0. \tag{8.24}$$

We note that since $(\overline{\tau} + w)(\overline{\tau} + w)^{\mathrm{T}} = E[(\overline{\tau}\overline{\tau}^{\mathrm{T}}) + w\overline{\tau}^{\mathrm{T}} + \overline{\tau}w^{\mathrm{T}} + ww^{\mathrm{T}}] \Rightarrow \overline{\tau}\overline{\tau}^{\mathrm{T}} + E(ww^{\mathrm{T}}) = \overline{\tau}\overline{\tau}^{\mathrm{T}} + \sigma^2I$. These two equations lead to the following result: $E[\text{trace}[\tilde{A}_1^{\mathrm{T}}\tilde{A}_1(\overline{\tau}\overline{\tau}^{\mathrm{T}} + \sigma^2I)]] = 0 = \text{trace}(\tilde{A}_1^{\mathrm{T}}\tilde{A}_1\overline{\tau}\overline{\tau}^{\mathrm{T}}) + \text{trace}(\tilde{A}_1^{\mathrm{T}}\tilde{A}_1\sigma^2I)$, which further implies $\|A_1\tau\|_F^2 + \|\tilde{A}_1\sigma\|_F^2 = 0$.

8.3.4 Optimality

This section shows the optimal solution of the original optimization problem can be arrived at by converging to the equilibrium point of the dynamic neural network (8.19).

(a)

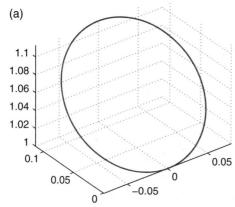

(b)

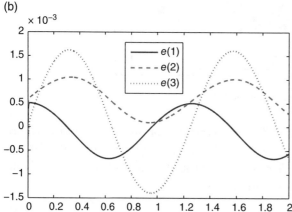

Figure 8.4 Tracking of a circular motion: (a) The tracking trajectory of the end-effector and (b) the time history of the position tracking error.

Theorem 8.1 For dynamic neural network (8.19) with $(\lambda_1^*, \lambda_2^*, \mu^*)$ as the equilibrium point, the output τ^* from (8.19d) is optimal to the constrained programming (8.18).

Proof: The equilibrium point $(\lambda_1^*, \lambda_2^*, \mu^*)$ meets the following condition according to state equations (8.19a):

$$-\Lambda_2^{-1}\lambda_1^* - A_1\Lambda_1^{-1}A_1^T\lambda_1^* - A_1\Lambda_1^{-1}A_2^T\lambda_2^* - \Lambda_2^{-1}B^T\mu^* = 0 \tag{8.25}$$

$$-A_2\Lambda^{-1}A_1^T\lambda_1^* - A_2\Lambda^{-1}A_2^T\lambda_2^* + \alpha = 0 \tag{8.26}$$

$$-\mu^* + (-B\Lambda_2^{-1}\lambda_1^* + \mu^* - B\Lambda_2^{-1}B^T\mu^* - b)^+ = 0 \tag{8.27}$$

and the corresponding output is:

$$\tau^* = A_1\Lambda_1^{-1}A_1^T\lambda_1^* + A_1\Lambda_1^{-1}A_2^T\lambda_2^*. \tag{8.28}$$

Define an auxiliary value,

$$\dot{\pi}^* = \Lambda^{-1}A_1^T\lambda_1^* + \Lambda^{-1}A_2^T\lambda_2^*. \tag{8.29}$$

τ^* is optimal to (8.18) can be obtained by showing the satirisation of Karush–Kuhn–Tucker conditions. Comparing the equation set composed of (8.25)–(8.28), and (8.29)

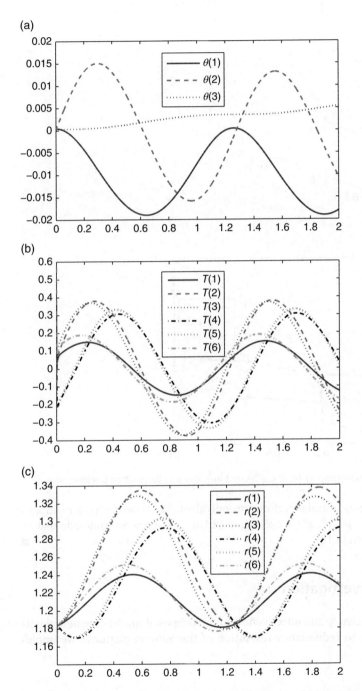

Figure 8.5 The time evolution of the Stewart platform state variables in the case of circular motion tracking. (a) Orientation of the platform θ; (b) control action τ; and (c) leg length r.

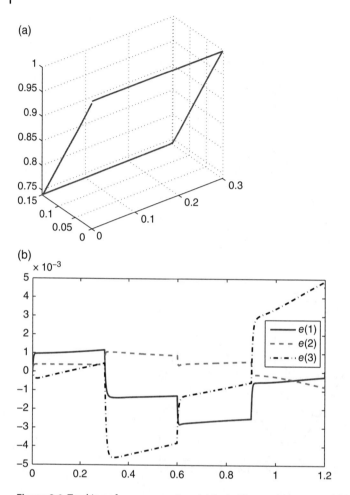

Figure 8.6 Tracking of a square motion. (a) End-effector trajectory and (b) position tracking error.

we find that they are identical and therefore are equivalent. The above procedure implies that the solution $(\lambda_1^*, \lambda_2^*, \mu^*, \tau^*, \dot{\pi}^*)$ is optimal to (8.18). Therefore, we conclude that τ^* is optimal to the problem (8.18). ■

8.4 Numerical Investigation

To demonstrate the efficiency and effectiveness of the proposed model-free neural network approach applied to redundancy resolution of the Stewart platform, we implemented it in MATLAB.

8.4.1 Setups

A Stewart platform with the following specifications will be considered: leg connectors are located around a circle with radius of 1.0 m at $b_1' = [0.7386, 0.1302, 0]$,

$b_2' = [0.7386, -0.1302, 0]$, $b_3' = [-0.4821, 0.5745, 0]$, $b_4' = [-0.2565, 0.7048, 0]$, $b_5' = [-0.2565, -0.7048, 0]$, and $b_6' = [-0.4821, 0.5745, 0]$ in the platform coordinate, and the leg connectors are located around the circle with radius of 0.75 m at $a_1 = [0.3750, 0.6495, 0]$, $a_2 = [0.3750, -0.6495, 0]$, $a_3 = [-0.7500, 0.0000, 0]$, $a_4 = [0.3750, 0.6495, 0]$, $a_5 = [0.3750, -0.6495, 0]$, and $a_6 = [-0.7500, 0.0000, 0]$ on the fixed base. For the ease of simulations the end-effector is rotated and placed at the origin with respect to the platform coordinate. The total expected redundancy is assumed to be 3, for the position tracking in the three-dimensional space. The input and output dimensions are 6 and 3, respectively.

The desired angular motion speed is set as 0.2 rad/s. The control scaling factor ε of the neural model is set as $\varepsilon = 10^{-2}$ and the learning scaling factor ζ is set as $\zeta = 10^{-4}$. The excitation signal w is set as random noise with zero mean and deviation of 10^{-3}. The basic idea is to set the noise at small value to ensure a minimal impact on the system performance.

In the simulation, we consider two tracking trajectories, namely, a square path and a circular path.

8.4.2 Circular Trajectory

In this section we simulate the tracking of a smooth circular path using the model-free approach. It is desired to follow the path of a circle at the minimal speed of 2 m/s. The circle is centered at $[-0.04, 0.06, 1.05]$ with a radius of 0.08 m, and has a revolutionary angle around the x axis for $45°$. In the simulation setup, Λ_1 and Λ_2 are chosen as the identity matrix. The value of the matrix A_1 is computed in real-time accordingly. The matrix A_2 is chosen as $A_2 = [I_{3\times3}, 0_{3\times3}]$ so that the position tracking requirements are obtained. In real time, the leg actuation speed is limited to a certain range as it is associated with the real-time constraints of the actuators.

The results are obtained by executing the simulation for 2 s. The Figure 8.4a shows the completed tracks in circular motion of an end-effector with the least tracking error as shown in Figure 8.4b. The position tracking error components $e(1)$, $e(2)$, and $e(3)$ are plotted along the x, y, and z axes of the base frame of the platform. The errors depicted in the figure are less than 0.015 m in amplitude. This path tracking simulation demonstrated the capability of the proposed model for resolving the kinematic redundancy of the physically constrained Stewart platform. The input motion for the legs is shown in Figure 8.5 which depicts the time evolution of the Stewart platform state variables, e.g. three Euler orientations of the platform, the position of the end-effector and the leg speed and its related length. The coordinates $p(1)$ and $p(2)$ of the attached moving frame start from zero and $p(3)$ varies between 0.005 and -0.006 m. Figure 8.5c states the bound ±0.25 ($\eta = 0.25$) for action speed. It can be observed that τ converges within the boundary region of $[-0.25, 0.25]$.

8.4.3 Square Trajectory

In this section we will discuss the simulated results of the square trajectory incorporating our model-free approach. In circular motion the path traveled by the end-effector in $360°$. However, in a square trajectory the path is nonsmooth switching from one straight line to the next, and how to reach timely control becomes a challenging issue.

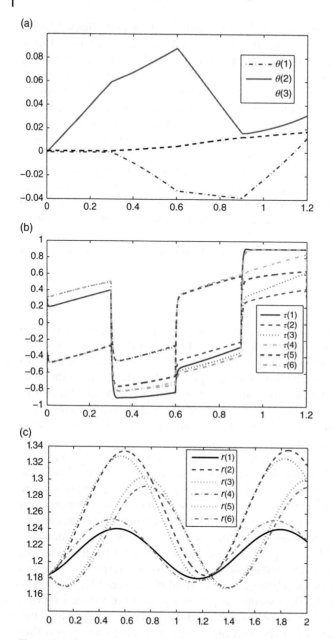

Figure 8.7 The time evolution of the platform state variables in the case of square motion tracking. (a) Orientation of the platform θ; (b) control action τ; and (c) leg length r.

The end-effector follows the trajectory which is centered at [0.15, 0.075, 0.74] with an edge length of 0.08 m, at a speed of 1.0 m/s. The revolution angle of the square around the x axis is 60°. The chosen parameters Λ_1 and Λ_2, A_2 and B have the same values as mentioned for the circular motion. The parameter A_1 is computed in real-time. The limit of speed $\eta = 0.6$ m/s and $b = 0.6\mathbf{I}_{12}$, and $\varepsilon = 0.001$. The tracking results for the square trajectory are shown in Figures 8.6 and 8.7 (position tracking error vector $[e(1), e(2), e(3)]$ is very small).

8.5 Summary

A model-free dual neural network is proposed to investigate and resolve the redundancy resolution problem of manipulators. In this chapter we also establish a dynamic model for a model-free network which is designed for a general case of a modular manipulator. The proposed controller model is designed such that a priori knowledge is not required for dynamic parameters and can suppress bounded external disturbances effectively in the presence of the external noise added. The instability problem caused by self-motions of the redundant robot can be resolved by the presented dual-neural control algorithm. Theoretical results are presented to verify the stability of the proposed models. The simulation is carried out on a redundant manipulator, which has verified the effectiveness of the dynamic modeling method and the controller design method.

Part III

Neural Networks for Cooperative Control

9

Zeroing Neural Networks for Robot Arm Motion Generation

9.1 Introduction

In recent decades, robotics has received more and more attention in scientific areas and engineering applications. Many research studies have focused on this topic, and various kinds of robots have been developed and investigated [1, 15, 63, 65–68, 98–101]. For example, a multiple collaborative manipulators system is investigated in [101], where the velocity consensus state can be achieved with the aid of the presented consensus control method. Redundant manipulators have been playing an increasingly important role and are an appealing topic in engineering fields [1, 15, 63, 98–100]. Redundant manipulators can achieve subtasks easily and dexterously such as repetitive motion and optimization of various performance criteria, since they possess more degrees of freedom than the minimum number required to execute a given primary task. One of the fundamental issues in operating such a robot system is the inverse-kinematics problem (also termed the redundancy-resolution problem). That is, with the desired trajectories of the manipulator's end-effector being provided in Cartesian space, trajectories in joint space should be generated accordingly [1, 15, 63, 98–100]. Rather than individual manipulators, multi-manipulator systems are used in many scenarios in order to improve performance and reliability. For example, in applications such as exploration, surveillance, tracking, or payload transport, in order to achieve better overall performance, a group of robots are driven to keep specific formations [101–105]. By considering the fact that manipulators often work in environments with strong electromagnetic interference, an unavoidable problem is to suppress noises during task execution. In addition to the increased complexity and computational burdens in multiple manipulator coordination, the noises involved also pose more requirements on real-time processing for the redundancy resolution at network level. Moreover, it is prior to design distributed algorithms requiring only local neighbor-to-neighbor information exchange because of limited communication bandwidth and communication range in many robotic applications.

Neural network approaches, recognized as a powerful tool for real-time processing, have long been applied widely in various control systems, e.g. multiple agents control [101, 106–111] and adaptive control [112–116]. For example, by using an adaptive neural network to approximate the nonlinear functions of the agents' dynamic, a consensus method based on the adaptive neural network is presented for the control of a class of

Kinematic Control of Redundant Robot Arms Using Neural Networks, First Edition.
Shuai Li, Long Jin and Mohammed Aquil Mirza.
© 2019 John Wiley & Sons Ltd. Published 2019 by John Wiley & Sons Ltd.

nonlinear second-order multi-agent systems, thereby greatly reducing the online computation burden. Liu and Tong construct an adaptive fuzzy controller for a class of nonlinear discrete-time systems with unknown functions and bounded disturbances in [109], which takes into account the effect of the discrete-time dead zone with the system states not being required to be measurable. Wang et al. present a control algorithm based on neural networks and backstepping techniques for the adaptive neural control of stochastic nonlinear systems in [114]. In [115], their model is further extended to be capable of controlling the stochastic systems with strong interconnected nonlinearities. The gradient-based neural network (GNN), being a typical class of the recurrent neural networks (RNNs), uses the norm of the error as the performance index and evolves along the gradient-descent direction to make the error norm vanish to zero in the time-invariant case [14]. As a special type of RNN designed for solving time-varying problems, the zeroing neural network (also termed the Zhang neural network, ZNN) is able to perfectly track a time-varying solution by using time derivatives of time-varying parameters [1, 14, 117–119]. Many researchers have made progress in this direction by proposing various kinds of ZNN models for solving problems with different highlights, e.g. finite-time convergence [120, 121] and high accuracy discretization [122, 123]. To generate repetitive motion in a closed path tracking, a drift-free scheme based on a ZNN-related neural-dynamic design method is investigated at the joint-velocity level in [124], which is further extended to schemes at joint-acceleration level in [14, 98, 99]. Although extensive achievements have been obtained for the control of single-arm redundant manipulators, research on the redundancy resolution of multiple manipulators is far from up-to-date, which severely restricts its applications in practical and academic research [125–127]. As mentioned previously, on account of their parallel computational power, RNNs are able to handle the problem of real-time control of manipulators. For example, a ZNN-based neural-dynamic design method has been exploited for the repetitive motion generation of dual-arm systems [63, 100], e.g. two arms of a humanoid robot [100].

In implementations of a neural network model for the cooperative motion generation of a network of redundant robot manipulators, we usually assume that it is free of all kinds of noises or external errors. However, there always exist some realization errors in hardware implementations or disturbances in the real-time control, which can be deemed as noises. Sometimes these noises have significant impacts on the accuracy of the neural network for solving theses time-varying problems, and in some cases, they may cause failure of the solving task [4]. In addition, for time critical tasks, it is preferable to integrate denoising with problem solving for real-time processing. That is, time is precious for time-varying problem solving (or the plant with time-varying demands) in practice, and preprocessing for noise reduction may consume extra time, possibly violating the requirement of real-time computation. A noise-tolerant ZNN (NTZNN) model is presented in [118] from the control perspective for solving time-varying problems, which can be employed for the cooperative motion generation of a network of redundant robot manipulators with inherent tolerance to noises. Comparisons between existing solutions for manipulator redundancy resolution and the proposed solution are summarized in Table 9.1. To the best of our knowledge, there is no systematic solution of noise-tolerant neural network design for the cooperative motion generation in a distributed network of redundant robot manipulators.

Table 9.1 Comparison of different schemes for redundancy resolution of manipulators.

	Noise tolerant	Manipulator numbers	Distributed vs. Centralized	Initial position
This chapter	Yes	Multiple	Distributed	Any
[1, 15, 98, 99]	No	Single	NA[a]	Any
[63, 100]	No	Two	Centralized	Any
[125]	No	Multiple	Distributed	Restrictive[b]
[126]	No	Multiple	Distributed	Restrictive[b]
[127]	No	Multiple	Distributed	Restrictive[b]

a) NA means that the item does not apply to the schemes in the associated references.
b) The associated schemes require that the initial position of the end-effector should be on the desired trajectory for tracking.

9.2 Preliminaries

In this section, definitions, assumptions, and modeling of the kinematics used in this chapter are provided to lay a foundation for investigations.

9.2.1 Problem Definition and Assumption

Definitions on the communication graph and communication topology presented in [105] are provided in the following.

A *communication graph* is a graph with the nodes being redundant manipulators and the edges being communication links. Moreover, $C(i)$ is used to denote a set of redundant manipulators with communication links to the ith redundant manipulator, which represents the neighbor set of the ith redundant manipulator on the communication graph.

Moveover, the definition of communication topology of limited communications is presented in the following [61, 105].

9.2.1.1 Assumption

Limited communication, with each redundant manipulator as a node and the communication link between one-hop neighboring robots as edges, is the communication topology of a connected undirected graph. We use $j \in \mathbb{N}(i)$ to denote the neighbor set of the ith redundant manipulator in the communication graph.

9.2.2 Manipulator Kinematics

Given the desired trajectory $r_d(t) \in \mathbb{R}^n$ of the end-effector, we want to generate online the joint trajectory $\vartheta(t) = [\vartheta_1(t), \vartheta_2(t), ..., \vartheta_m(t)]^T \in \mathbb{R}^m$ so as to command the manipulator motion. Note that the Cartesian coordinate $r \in \mathbb{R}^n$ in the workspace of a manipulator is uniquely determined by a nonlinear mapping:

$$r(t) = f(\vartheta(t)), \tag{9.1}$$

where $f(\cdot)$ is a differentiable nonlinear function with a known structure and parameters for a given manipulator. Computing time derivations on both sides of (9.1) leads to

$$\dot{r}(t) = J(\vartheta(t))\dot{\vartheta}(t), \tag{9.2}$$

where $J(\vartheta(t)) = \partial f / \partial \vartheta \in \mathbb{R}^{n \times m}$ is the Jacobian matrix of $f(\vartheta(t))$, and usually is abbreviated as J. The end-effector $r(t)$ of the redundant manipulator is expected to track the desired path $r_d(t)$, i.e. $r(t) \to r_d(t)$.

9.3 Problem Formulation and Distributed Scheme

In this section, the problem formulation is constructed and then the associated distributed scheme is presented.

9.3.1 Problem Formulation and Neural-Dynamic Design

During the executing process of the cooperative motion generation of redundant robot manipulators, the desired formation of all end-effectors should be maintained to avoid stretching or squeezing of the payload generated by the relative movement between them. In this chapter, the commanding signal is only accessible to its neighboring manipulators, instead of all manipulators. Therefore, the constraint for the ith manipulator can be formulated as

$$\sum_{j \in C(i)} A_{ij}(\kappa_i(t) - \kappa_j(t)) + \delta_i(\kappa_i(t) - r_d(t)) = 0, \tag{9.3}$$

where $C(i)$ denotes the neighbor set of the ith manipulator on the communication graph; A_{ij} denotes the connection weight between the ith manipulator and the jth one, with its value being 1 for $j \in C(i)$ and 0 for $j \notin C(i)$; $\kappa_i(t) = r_i(t) - r_{rp}$ with r_{rp} denoting the constant distance vector between the end-effector and the reference point; and $\delta_i = 1$ for $i \in C(0)$ and $\delta_i = 0$ for $i \notin C(0)$ with $C(0)$ denoting the neighbor set of the command center. By combining the constraints of all manipulators in the group, the constraint for cooperative control of redundant robot manipulators can be formulated as

$$(L \otimes I_n)\overline{\kappa}(t) + (\Lambda \otimes I_n)(\overline{\kappa}(t) - \mathbf{1}_p \otimes r_d(t)) = 0, \tag{9.4}$$

where $\otimes$ is the Kronecker product; $\mathbf{1}_p \in \mathbb{R}^{p \times 1}$ denotes a vector composed of 1; $I_n \in \mathbb{R}^{n \times n}$ is an identity matrix; Laplacian matrix $L = \text{diag}(A\mathbf{1}_p) - A \in \mathbb{R}^{p \times p}$ with $\text{diag}(A\mathbf{1}_p)$ being the diagonal matrix whose p diagonal entries are the p elements of the vector $A\mathbf{1}_p \in \mathbb{R}^{p \times 1}$ with the ijth element of matrix A being A_{ij}; $\overline{\kappa}(t) = [\kappa_1^T(t), \cdots, \kappa_p^T(t)]^T \in \mathbb{R}^{np \times 1}$; and $\Lambda \in \mathbb{R}^{p \times p}$ is a diagonal matrix defined as

$$\Lambda_{ij} = \begin{cases} \delta_i, & \text{if } i = j, \\ 0, & \text{if } i \neq j. \end{cases}$$

A vector-valued error function could be defined to construct the Jacobian equality constraint:

$$\mu(t) = (L \otimes I_n)\overline{\kappa}(t) + (\Lambda \otimes I_n)(\overline{\kappa}(t) - \mathbf{1}_p \otimes r_d(t)). \tag{9.5}$$

By exploiting the neural-dynamic method presented in [115], we have

$$\dot{\mu}(t) = \gamma\mu(t), \tag{9.6}$$

where $\gamma > 0$. Expanding (9.6) leads to

$$(L + \Lambda) \otimes \bar{J}(\vartheta)\dot{\bar{\vartheta}}(t) = \Lambda \otimes I_n \cdot \mathbf{1}_p \otimes \dot{r}_{\mathrm{d}}(t) - \gamma((L \otimes I_n)\bar{\kappa}(t)$$
$$+ (\Lambda \otimes I_n)(\bar{\kappa}(t) - \mathbf{1}_p \otimes r_{\mathrm{d}}(t))), \tag{9.7}$$

where $\bar{\vartheta}(t) = [\vartheta_1^{\mathrm{T}}(t), \cdots, \vartheta_p^{\mathrm{T}}(t)]^{\mathrm{T}} \in \mathbb{R}^{mp \times 1}$ with $\dot{\bar{\vartheta}}(t)$ denoting its time derivative; and

$$\bar{J}(\bar{\vartheta}) = \begin{bmatrix} J_1(\vartheta_1) & 0 & \cdots & 0 \\ 0 & J_2(\vartheta_2) & \cdots & 0 \\ \vdots & \vdots & \ddots & \vdots \\ 0 & 0 & \cdots & J_p(\vartheta_p) \end{bmatrix} \in \mathbb{R}^{np \times mp}.$$

9.3.2 Distributed Scheme

We consider the following minimum velocity norm (MVN) performance index in this chapter:

$$U = \sum_{i=1}^{p} \dot{\vartheta}_i^{\mathrm{T}}(t)\dot{\vartheta}_i(t)/2 = \parallel \dot{\bar{\vartheta}}(t) \parallel_2^2 /2.$$

Then, the distributed scheme is formulated as

$$\min \quad \parallel \dot{\bar{\vartheta}}(t) \parallel_2^2 /2, \tag{9.8}$$
$$\text{s.t.} \quad (L + \Lambda) \otimes \bar{J}(\vartheta)\dot{\bar{\vartheta}}(t) = \Lambda \otimes I_n \cdot \mathbf{1}_p \otimes \dot{r}_{\mathrm{d}}(t) - \gamma((L \otimes I_n)\bar{\kappa}(t)$$
$$+ (\Lambda \otimes I_n)(\bar{\kappa}(t) - \mathbf{1}_p \otimes r_{\mathrm{d}}(t))).$$

So far, the MVN-oriented distributed scheme has been formulated as a quadratic programming problem with limited communications. In the following section, the NTZNN model will be constructed to solve such a problem online. Note that the differences of different types of QP-based redundancy-resolution schemes (e.g. MVN-type scheme, RMP-type scheme, or even their weighted combinations) lie in the performance index. Therefore, the proposed distributed scheme is capable of other performance indices and can be extended by following similar steps to those presented in this chapter.

9.4 NTZNN Solver and Theoretical Analyses

Here, we present an NTZNN model to solve the MVN-oriented distributed scheme (9.8) for the motion generation of multiple redundant robot manipulators with guaranteed convergence and robustness.

9.4.1 ZNN for Real-Time Redundancy Resolution

As a novel type of RNN specifically designed from the perspective of control for solving time-varying problems, NTZNN is able to accurately track a time-varying solution with the guaranteed capability of noise tolerance and suppression. Therefore, we exploit an NTZNN model for the online solution of the MVN-oriented distributed scheme (9.8).

Define a Lagrange function as follows:

$$
\begin{aligned}
\Upsilon(\bar{\vartheta}(t), \xi(t)) = {}& \bar{\dot{\vartheta}}^{\mathrm{T}}(t)\bar{\dot{\vartheta}}(t)/2 \\
& + \xi^{\mathrm{T}}(t)((L+\Lambda) \otimes \bar{J}(\vartheta)\bar{\dot{\vartheta}}(t) - \Lambda \otimes I_n \cdot \mathbf{1}_p \otimes \dot{r}_\mathrm{d}(t) \\
& + \gamma((L \otimes I_n)\overline{\kappa}(t) + (\Lambda \otimes I_n)(\overline{\kappa}(t) - \mathbf{1}_p \otimes r_\mathrm{d}(t)))),
\end{aligned}
$$

where $\xi \mathbf{1}_p \otimes \dot{r}_\mathrm{d}(t) \in \mathbb{R}^{np}$ is the Lagrange-multiplier vector. According to the Lagrange-multiplier method [58], the solution has to satisfy the following

$$
\frac{\partial \Upsilon(\bar{\vartheta}(t), \xi(t))}{\partial \bar{\vartheta}(t)} = 0, \quad \frac{\partial \Upsilon(\bar{\vartheta}(t), \xi(t))}{\partial \xi(t)} = 0. \tag{9.9}
$$

Then, solving (9.8) can be done by solving the following equation:

$$
\Theta(t)x(t) = \eta(t), \tag{9.10}
$$

where

$$
\Theta(t) = \begin{bmatrix} I_{mp} & ((L+\Lambda) \otimes \bar{J}(\bar{\vartheta}))^{\mathrm{T}} \\ (L+\Lambda) \otimes \bar{J}(\bar{\vartheta}) & 0 \end{bmatrix},
$$

$$
x(t) = \begin{bmatrix} \bar{\dot{\vartheta}}(t) \\ \xi(t) \end{bmatrix}, \quad \eta(t) = \begin{bmatrix} 0 \\ u(t) \end{bmatrix},
$$

with $u = \Lambda \otimes I_n \cdot \mathbf{1}_p \otimes \dot{r}_\mathrm{d} - \gamma((L \otimes I_n)\overline{\kappa} + (\Lambda \otimes I_n)(\overline{\kappa} - \mathbf{1}_p \otimes r_\mathrm{d}))$. We define the following vector-valued indefinite error function as

$$
\zeta(t) = \Theta(t)x(t) - \eta(t). \tag{9.11}
$$

From a control perspective, the error $\zeta(t)$ can be interpreted as a distance measure between $x(t)$ and $x^*(t)$. If the error $\zeta(t)$ approaches zero with time, then the variable $x(t)$ is correspondingly driven to $x^*(t)$ of (9.11). To force $\zeta(t)$ to be zero, the following NTZNN design formula is adopted:

$$
\dot{\zeta}(t) = -\varepsilon_1 \zeta(t) - \varepsilon_2 \int_0^t \zeta(\tau)\mathrm{d}\tau, \tag{9.12}
$$

where $\varepsilon_1 > 0$ and $\varepsilon_2 > 0$. We further obtain the following distributed NTZNN model:

$$
\begin{aligned}
\Theta(t)\dot{x}(t) = {}& -\dot{\Theta}(t)x(t) - \varepsilon_1\left(\Theta(t)x(t) - \eta(t)\right) \\
& - \varepsilon_2 \int_0^t (\Theta(\tau)x(\tau) - \eta(\tau))\mathrm{d}\tau + \dot{\eta}(t). \tag{9.13}
\end{aligned}
$$

In addition, the architecture for an electronic implementation of the distributed NTZNN model (9.13) is depicted in Figure 9.1.

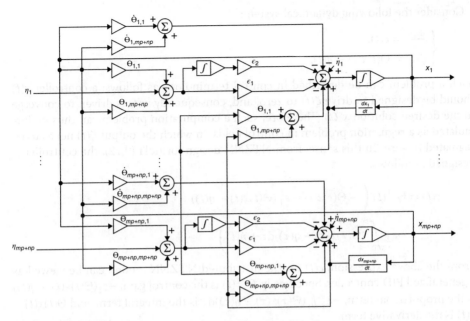

Figure 9.1 Circuit schematic of the distributed NTZNN model (9.13).

To lay a basis for further investigation on the robustness of the MVN-oriented distributed scheme (9.8) aided by distributed NTZNN model (9.13) under the pollution of unknown noises, we have the following equation:

$$\Theta(t)\dot{x}(t) = -\dot{\Theta}(t)x(t) - \varepsilon_1\left(\Theta(t)x(t) - \eta(t)\right)$$

$$- \varepsilon_2 \int_0^t (\Theta(\tau)x(\tau) - \eta(\tau))\mathrm{d}\tau + \dot{\eta}(t) + \rho(t), \tag{9.14}$$

where $\rho(t) \in \mathbb{R}^{np+mp}$ denotes the vector-form noises.

Remark 9.1 One major contribution of this work can be viewed as a nonlinear dynamic transformation, which converts a nonlinear dynamic system into a regular second-order linear dynamic system. For instance, there intrinsically exists nonlinear dynamics in the robot manipulator and it is not a second-order linear dynamic system. However, our approach can effectively simplify and solve the MVN-oriented motion generation problem of it by mapping the nonlinearity in the original dynamics into a second order linear one. In addition, researchers would generally think that the distributed NTZNN model and traditional PID controllers should be the same. However, analysis shows that it should be another case, which is presented in the following remark.

Remark 9.2 We may think that $\mathrm{d}(\Theta(t)x(t) - \eta(t))/\mathrm{d}t = \dot{\Theta}(t)x(t) + \Theta(t)\dot{x}(t) - \dot{\eta}(t)$ should serve as the derivative term in analogy to the PID control. However, counter-intuitively, the distributed NTZNN design methodology reveals that it is $\dot{\Theta}(t)x(t) - \dot{\eta}(t)$ without the term $\Theta(t)\dot{x}(t)$ as the derivative term that can guarantee the overall convergence and robustness.

Consider the following dynamical system:

$$\begin{cases} \frac{dx(t)}{dt} = c(t), \\ \zeta(t) = \Theta(t)x(t) - \eta(t). \end{cases}$$

Such a problem can be described in control terminology as follows: a controller $c(t)$ should be designed to drive $\zeta(t)$ to zero, and, consequently, $x(t)$ is driven to converge to the desired solution $x^*(t)$. Therefore, such a computation problem can thus be formulated as a regulation problem in control fields, in which the output $\zeta(t)$ needs to be regulated to zero. In this sense, from NTZNN design formula (9.12), the controller is designed as follows:

$$c(t) = \Theta^{-1}(t)\left(-\dot{\Theta}(t)x(t) - \varepsilon_1\left(\Theta(t)x(t) - \eta(t)\right) \right.$$
$$\left. -\varepsilon_2 \int_0^t (\Theta(\tau)x(\tau) - \eta(\tau))d\tau + \dot{\eta}(t) \right).$$

From the above expression, $c(t)$ (or the distributed NTZNN model) can be viewed as a generalized PID controller. Specifically, $\Theta^{-1}(t)$ is the control gain, $-\varepsilon_1(\Theta(t)x(t) - \eta(t))$ is the proportional term, $-\varepsilon_2 \int_0^t (\Theta(\tau)x(\tau) - \eta(\tau))d\tau$ is the integral term, and $\dot{\Theta}(t)x(t) - \dot{\eta}(t)$ is the derivative term.

Remark 9.3   The control diagram of the MVN-oriented distributed scheme (9.8) aided by distributed NTZNN model (9.13) for the cooperative motion generation of a network of redundant robot manipulators is shown in Figure 9.2. It can be observed from this figure that, at each time instant, the NTZNN model (9.13) receives the desired information $r_d, \dot{r}_d, \ddot{r}_d$ and the feedback information $\bar{\kappa}$ and then outputs the control input $\bar{\vartheta}$ to drive all the robot manipulators to execute the given task.

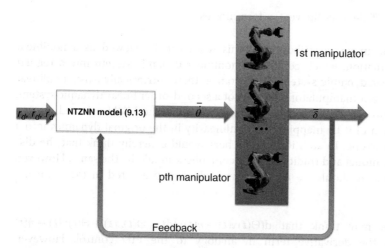

Figure 9.2 Control diagram of the MVN-oriented distributed scheme (9.8) aided by the distributed NTZNN model (9.13) for the cooperative motion generation of a network of redundant robot manipulators.

9.4.2 Theoretical Analyses and Results

The analyses on the stability, convergence and robustness of the proposed distributed NTZNN model (9.13) are conducted via the following theorems.

Theorem 9.1 Distributed NTZNN model (9.13) is stable and is exponentially and globally convergent to an equilibrium point x^*, of which the first np elements constitute the optimal solution $\overrightarrow{\vartheta}^*$ to the cooperative motion generation of redundant robot manipulators synthesized by the MVN-oriented distributed scheme (9.8).

Proof: In terms of solving the MVN-oriented distributed scheme (9.8), the distributed NTZNN model (9.13) is an equivalent expansion of the design formula $\dot{\zeta}(t) = -\varepsilon_1 \zeta(t) - \varepsilon_2 \int_0^t \zeta(\tau)\mathrm{d}\tau$ with $\zeta(t) = \Theta(t)x(t) - \eta(t)$. Therefore, the design formula is directly used in this proof. By selecting a Lyapunov function candidate $v(t) = \zeta^2(t) + \varepsilon_2(\int_0^t \zeta(\tau)\mathrm{d}\tau)^2$ and by using the Lyapunov theory, one can readily derive that the distributed NTZNN model (9.13) is stable with $\dot{v}(t) = -2\varepsilon_1 \zeta^2(t)$.

In addition, solving the second-order linear dynamic system $\dot{\zeta}(t) = -\varepsilon_1 \zeta(t) - \varepsilon_2 \int_0^t \zeta(\tau)\mathrm{d}\tau$ directly, we have the conclusion that, starting from any initial condition $\zeta(0)$, the residual error $\zeta(t)$ of the distributed NTZNN model (9.13) for solving the MVN-oriented distributed scheme (9.8) globally and exponentially converges to zero. That is, the state vector $x(t)$ of the distributed NTZNN model (9.13) globally and exponentially converges to an equilibrium point x^*, of which the first np elements constitute the optimal solution $\overrightarrow{\vartheta}^*$ to the cooperative motion generation of redundant robot manipulators synthesized by MVN-oriented distributed scheme (9.8). The proof is thus completed. ∎

There always exist communication noises, computational errors, perturbations or even their superpositions for the cooperative motion generation of redundant robot manipulators, which can be deemed as noises. Sometimes those noises have significant impacts on the accuracy of the computational model for the problem solving, and in some cases, they may cause failure of the solving task. Therefore, it is worth investigating the robustness of the distributed NTZNN model (9.13). The following theorem is presented to discuss the performance of the distributed NTZNN model (9.13) with constant noise $\rho(t) = \varsigma$.

Theorem 9.2 Distributed NTZNN model (9.13) is stable and is globally convergent to an equilibrium point x^* no matter how large the unknown constant noise $\rho(t) = \varsigma$ is. Specifically, the vector constituted by the first np elements constitute the optimal solution $\overrightarrow{\vartheta}^*$ to the cooperative motion generation of redundant robot manipulators synthesized by the MVN-oriented distributed scheme (9.8).

Proof: The noise-polluted distributed NTZNN model (9.14) is an equivalent expansion of $\dot{\zeta}(t) = -\varepsilon_1 \zeta(t) - \varepsilon_2 \int_0^t \zeta(\tau)\mathrm{d}\tau + \varsigma$. Using Laplace transform to its ith subsystem leads to

$$s\zeta_i(s) - \zeta_i(0) = -\varepsilon_1 \zeta_i(s) - \frac{\varepsilon_2}{s}\zeta_i(s) + \varsigma, \tag{9.15}$$

with the transfer function being $s/(s^2 + \varepsilon_1 s + \varepsilon_2)$. It can be readily deduced that such a system is stable in view of the fact that the poles locate on the left half-plane. Then, using the final value theorem to (9.15), we have

$$\lim_{t \to \infty} \zeta_i(t) = \lim_{s \to 0} s\zeta_i(s) = \lim_{s \to 0} \frac{s^2(\zeta_i(0) + \varsigma_i/s)}{s^2 + \varepsilon_1 s + \varepsilon_2} = 0.$$

Therefore, it can be concluded from (9.7) that, for $t \to \infty$,

$$((L + \Lambda) \otimes \bar{J}(\bar{\vartheta})\bar{\vartheta}(t))_i - (\Lambda \otimes I_n \cdot \mathbf{1}_p \otimes \dot{r}_d(t)$$
$$-\gamma((L \otimes I_n)\bar{\kappa}(t) + (\Lambda \otimes I_n)(\bar{\kappa}(t) - \mathbf{1}_p \otimes r_d(t))))_i$$
$$= 0,$$

which can be further rewritten as

$$\dot{\mu}_i(t) = \gamma \mu_i(t) + 0.$$

Using Laplace transform to the above subsystem and then applying the final value theorem to it leads to

$$\lim_{t \to \infty} \mu_i(t) = \lim_{s \to 0} s\mu_i(s) = 0.$$

Therefore, we have the conclusion that distributed NTZNN model (9.13) is stable and is globally convergent to an equilibrium point x^* no matter how large the unknown constant noise $\rho(t) = \varsigma$ is. Specifically, the vector constituted by the first np elements constitute the optimal solution $\bar{\vartheta}^*$ to the cooperative motion generation of redundant robot manipulators synthesized by the MVN-oriented distributed scheme (9.8). ∎

In addition, regarding the random noise, we have the following theorem.

Theorem 9.3 Perturbed with bounded unknown random noise $\rho(t) = \phi(t) \in \mathbb{R}^{mp+np}$, the cooperative control of redundant robot manipulators synthesized by MVN-oriented distributed scheme (9.8) and solved by distributed NTZNN model (9.13) can achieve arbitrary accuracy with sufficiently large γ, sufficiently large ε_1, and with a proper ε_2.

Proof: By following similar steps to those in Theorem 9.2, we can write the ith subsystem of bounded unknown random noise $\rho(t) = \phi(t) \in \mathbb{R}^{mp+np}$ polluted distributed NTZNN model (9.14) as

$$\dot{\zeta}_i(t) = -\varepsilon_1 \zeta_i(t) - \varepsilon_2 \int_0^t \zeta_i(\tau)d\tau + \phi_i(t).$$

Depending on the values of ε_1 and ε_2, the analyses can be divided into the following three situations.

- For $\varepsilon_1^2 > 4\varepsilon_2$, the solution can be obtained as

$$\zeta_i(t) = \frac{\zeta_i(0)(\iota_1 \exp(\iota_1 t) - \iota_2 \exp(\iota_2 t))}{(\iota_1 - \iota_2)}$$
$$+ \left(\int_0^t (\iota_1 \exp(\iota_1(t - \tau)) - \iota_2 \exp(\iota_2(t - \tau)))\phi_i(\tau)d\tau \right) \frac{1}{(\iota_1 - \iota_2)},$$

where $\iota_{1,2} = \left(-\varepsilon_1 \pm \sqrt{\varepsilon_1^2 - 4\varepsilon_2}\right)/2$. Based on the triangle inequality, we have

$$|\zeta_i(t)| \le \frac{|\zeta_i(0)(\iota_1 \exp(\iota_1 t) - \iota_2 \exp(\iota_2 t))|}{(\iota_1 - \iota_2)}$$

$$+ \frac{\int_0^t |\iota_1 \exp(\iota_1(t - \tau))||\phi_i(\tau)| d\tau}{(\iota_1 - \iota_2)}$$

$$+ \frac{\int_0^t |\iota_2 \exp(\iota_2(t - \tau))||\phi_i(\tau)| d\tau}{(\iota_1 - \iota_2)}.$$

We further have

$$|\zeta_i(t)| \le \frac{|\zeta_i(0)(\iota_1 \exp(\iota_1 t) - \iota_2 \exp(\iota_2 t))|}{(\iota_1 - \iota_2)}$$

$$+ \frac{2}{(\iota_1 - \iota_2)} \max_{0 \le \tau \le t} |\phi_i(\tau)|$$

$$= \frac{|\zeta_i(0)(\iota_1 \exp(\iota_1 t) - \iota_2 \exp(\iota_2 t))|}{(\iota_1 - \iota_2)}$$

$$+ \frac{2}{\sqrt{\varepsilon_1^2 - 4\varepsilon_2}} \max_{0 \le \tau \le t} |\phi_i(\tau)|.$$

Finally, we have

$$\limsup_{t \to \infty} \| \zeta(t) \|_2 \le \frac{2\tilde{\phi}\sqrt{n + m}}{\sqrt{\varepsilon_1^2 - 4\varepsilon_2}},$$

with $\tilde{\phi} = \max_{1 \le i \le n+m}\{\max_{0 \le \tau \le t} |\phi_i(\tau)|\}$.

- For $\varepsilon_1^2 = 4\varepsilon_2$, according to the proof of Theorem 1 in [14], there exist $v_1 > 0$ and $v_2 > 0$, such that

$$|\iota_1|t \exp(\iota_1 t) \le v_1 \exp(-v_2 t).$$

Thus, based on the above inequality and following similar steps to the situation of $\varepsilon_1^2 > 4\varepsilon_2$, we have

$$\limsup_{t \to \infty} \| \zeta(t) \|_2 \le \left(\frac{v_1}{v_2} - \frac{1}{\iota_1}\right) \tilde{\phi}\sqrt{n + m},$$

with $\tilde{\phi} = \max_{1 \le i \le n+m}\{\max_{0 \le \tau \le t} |\phi_i(\tau)|\}$.

- For $\varepsilon_1^2 < 4\varepsilon_2$, following similar steps to the situation of $\varepsilon_1^2 > 4\varepsilon_2$, we have

$$\limsup_{t \to \infty} \| \zeta(t) \|_2 \le \frac{4\varepsilon_2 \tilde{\phi}\sqrt{n + m}}{\varepsilon_1\sqrt{4\varepsilon_2 - \varepsilon_1^2}},$$

with $\tilde{\phi} = \max_{1 \le i \le n+m}\{\max_{0 \le \tau \le t} |\phi_i(\tau)|\}$.

Letting $\rho = \lim_{t\to\infty} \sup \| \zeta(t)\|_2$, following the similar steps presented in Theorem 9.2, we have the conclusion that

$$\lim_{t\to\infty} \| \mu(t)\|_2 \leq \frac{\rho}{\gamma}.$$

Therefore, we have the conclusion that, perturbed with bounded unknown random noise $\rho(t) = \phi(t) \in \mathbb{R}^{mp+np}$, the cooperative motion generation of redundant robot manipulators synthesized by MVN-oriented distributed scheme (9.8) and solved by distributed NTZNN model (9.13), can achieve arbitrary accuracy with sufficiently large γ, sufficiently large ε_1, and with a proper ε_2. The proof is thus completed. ∎

9.5 Illustrative Examples

There are two procedures for a distributed network of redundant manipulators cooperatively transporting a payload. The first procedure: each end-effector of all manipulators should reach a desired configuration (the consensus of redundant robot manipulators) for holding a different place or a handle on the payload. The second procedure: each end-effector of all manipulators should keep a desired format during the execution to fulfill certain tasks (the cooperative control of redundant robot manipulators). That is to say, a reference point on the payload, e.g. the center of mass, is expected to stay at the reference position or track the reference trajectory. In this section, computer simulations on the motion generation (consensus and control) of redundant robot manipulators are conducted based on a group of PUMA 560 manipulators to illustrate the effectiveness of the proposed MVN-oriented distributed scheme (9.8) as well as the distributed NTZNN model (9.13). The PUMA 560 robot manipulator has six revolutionary joints, which connect seven links in series with the last one being an end-effector. In general, when we only consider the position of the end-effector, a PUMA 560 robot manipulator can be deemed as a functionally redundant robot. For the modeling details on the kinematics of a PUMA 560 robot manipulator, please refer to [125].

9.5.1 Consensus to a Fixed Configuration

In this consensus example, we choose $\gamma = 10$, $\varepsilon_1 = 10$, $\varepsilon_2 = 10$, $p = 8$, constant noise $\rho(t) = 10$, and task duration is 5 s; the initial joint state of each manipulator is randomly generated. In addition, A_{ij} is set as

$$A_{ij} = \begin{cases} 1, & \text{if } |i - j| \leq 2 \\ 0, & \text{otherwise,} \end{cases}$$

and $\delta_1 = \delta_5 = 1$ with $\delta_i = 0$ for $i = 2, 3, 4, 6, 7, 8$. Besides, the initial values of distributed NTZNN model (9.13) are set as zero (e.g. $\bar{\xi}$). The corresponding simulation results based on the MVN-oriented distributed scheme (9.8) for consensus of redundant robot manipulators solved by distributed NTZNN model (9.13) are illustrated in Figure 9.3.

Specifically, Figure 9.3b plots that trajectories of end-effectors of eight PUMA 560 manipulators are all driven to the desired positions. Figure 9.3b shows that, after the transient state, the position errors of the end-effector converge to zero rapidly in spite of noises. In addition, Figure 9.3c and d illustrate the smooth profiles of joint angle and

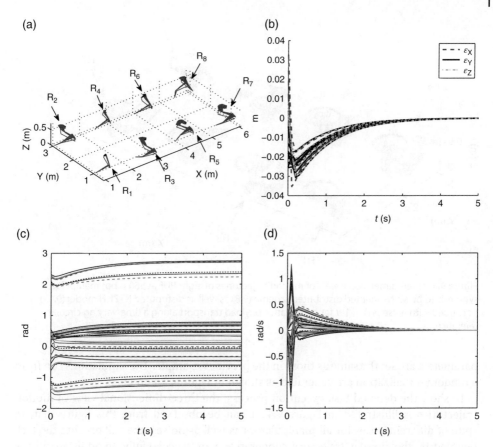

Figure 9.3 Computer simulations synthesized by MVN-oriented distributed scheme (9.8) and distributed NTZNN model (9.13) for consensus of eight redundant PUMA 560 robot manipulators with limited communications and perturbed with constant noise $\rho(t) = 10$, where initial joint states of manipulators are randomly generated and R_i denotes the ith redundant robot manipulator. (a) Motion trajectories; (b) profiles of end-effector position errors; (c) profiles of joint angle; and (d) profiles of joint velocity.

joint velocity, respectively, which further demonstrates the effectiveness of the proposed MVN-oriented distributed scheme (9.8) for the consensus of multiple redundant manipulators in the presence of noises.

9.5.2 Cooperative Motion Generation Perturbed by Noises

In this section, we consider eight PUMA 560 manipulators for cooperative payload transport perturbed with noise $\rho(t) = 10t$. The desired motion issued by the command center is to track a circular path with radius 0.3 m at an angular speed of 0.2 rad/s around its center. In addition, only manipulators 1 and 5 are able to access the desired motion information provided by the command center. The parameters are chosen as $\gamma = 10$, $\varepsilon_1 = 100$, $\varepsilon_2 = 100$, and task duration is $2\pi/0.2 = 31.4$ s. In addition, the rest

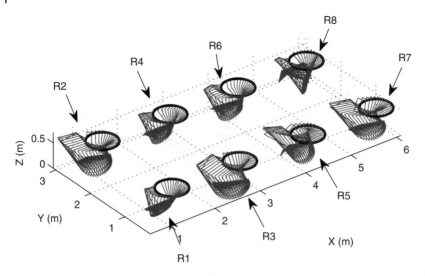

Figure 9.4 Three-dimensional view of motion trajectories of eight PUMA 560 robot manipulators synthesized by MVN-oriented distributed scheme (9.8) as well as distributed NTZNN model (9.13) perturbed with noise $\rho(t) = 10t$ for cooperative payload transport along a time-varying circular reference.

parameters are set the same as those in the previous example. The simulations run from a random initialization are generated as shown in Figures 9.4 and 9.5.

To show the detailed task execution process, the three-dimensional view of motion trajectories is illustrated in Figure 9.4. It can be observed from this figure that, in spite of different poses for all manipulators as well as noises, they all execute the task denoted by the circular trajectory cooperatively and successfully. In addition, all the end-effector position tracking errors shown in Figure 9.5a are of order 10^{-4} m, which remain at a very tiny value during the task execution. The corresponding profiles of joint angle and joint velocity are shown in Figure 9.5c and d, respectively. It is observable in Figure 9.5c that the resulting $\bar{\vartheta}$ move smoothly. These simulation results substantiate the effectiveness of MVN-oriented distributed scheme (9.8) and distributed NTZNN model (9.13).

9.5.3 ZNN-Based Solution Perturbed by Noises

In this section, we show the cooperative motion generation with the aid of the original ZNN model perturbed by noises. A ZNN-based solution is produced when $\varepsilon_2 = 0$ with the rest of the parameters set the same as those in the previous example. It can be observed from Figure 9.6a that, perturbed with noise $\rho(t) = 10t$, the end-effector position errors are larger than those illustrated in Figure 9.5a. In addition, it can be found in Figure 9.6b and c that the corresponding joint-angle and joint-velocity profiles oscillate with the time, which further substantiates the superiority and effectiveness of MVN-oriented distributed scheme (9.8) with the aid of distributed NTZNN model (9.13).

Figure 9.5 Computer simulations synthesized by MVN-oriented distributed scheme (9.8) as well as distributed NTZNN model (9.13) perturbed with noise $\rho(t) = 10t$ for cooperative payload transport of eight PUMA 560 robot manipulators along a time-varying circular reference with limited communications. Profiles of (a) position errors, (b) joint angle, and (c) joint velocity.

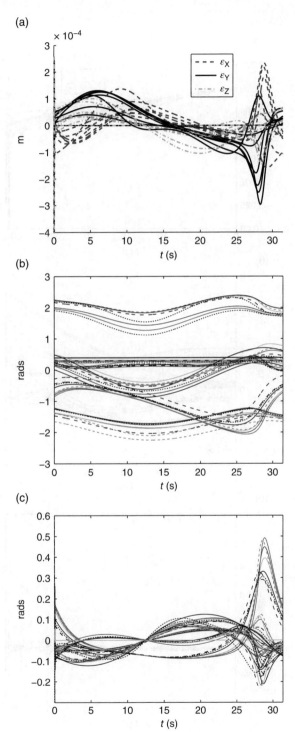

(a)

(b)

(c)

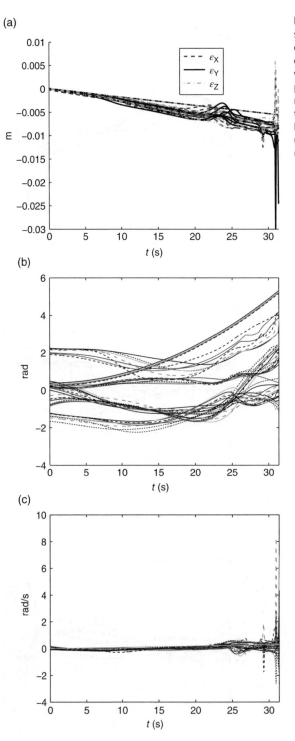

Figure 9.6 Computer simulations synthesized by MVN-oriented distributed scheme (9.8) as well as distributed ZNN model perturbed with noise $\rho(t) = 10t$ for cooperative payload transport of eight PUMA 560 robot manipulators along a time-varying circular reference with limited communications. Profiles of (a) end-effector position errors, (b) joint angle, and (c) joint velocity.

9.6 Summary

In this chapter, a distributed scheme has been proposed for the cooperative motion generation of multiple redundant manipulators. The proposed scheme can simultaneously achieve the specified primary task to reach global cooperation under limited communications among manipulators and optimality in terms of a specified optimization index of redundant robot manipulators. The proposed distributed scheme has been reformulated as a quadratic program. To inherently suppress noises, a NTZNN has been constructed to solve the quadratic program problem online. Then, theoretical analyses have shown that, without noise, the proposed distributed scheme is able to execute a given task with exponentially convergent position errors. Moreover, in the presence of noise, the proposed distributed scheme with the aid of a NTZNN model has a satisfactory performance. Furthermore, simulations and comparisons based on PUMA 560 redundant robot manipulators have substantiated the effectiveness and accuracy of the proposed distributed scheme with the aid of a NTZNN model.

10

Zeroing Neural Networks for Robot Arm Motion Generation

10.1 Introduction

The model for solving nonlinear equations bears an essential similarity with the controller for controlling the plant: their residual errors are required to decrease to an acceptable small value as soon as possible. The exploitation of this similarity provides a possibility to investigate computational methods from the perspective of control system theory. It is worth mentioning that, in the field of numerical computation and control, there is always a great demand for robustness due to the existence of various noises or disturbances, such as round-off errors and truncation errors. From the perspective of computation, many recurrent neural network models, e.g. zeroing neural network (ZNN), are analyzed and applied to the solution of various problems [128]. To improve the robustness for solving time-varying problems, a noise-tolerant zeroing neural network (NTZNN) design formula is proposed in [13], which can be used to design recurrent neural networks from the viewpoint of control. Then, such a NTZNN design method is explored to design a modified NTZNN in [1] for the online solution of quadratic programming with application to the repetitive motion planning of a redundant robot arm. As discussed in [7], nonlinear activation functions can be used to accelerate the convergence speed of original ZNN models. However, to the best of the authors' knowledge, there is no systematic solution on NTZNN with the aid of nonlinear activation functions. Therefore, the extension from existing NTZNN to nonlinearly activated NTZNN (NANTZNN) remains an unsolved problem.

The redundant robot arms are robotic devices, of which the available degrees of freedom are more than those strictly required for executing the user-specified primary end-effector task [15]. Recent progress in this topic shows the advantages of using optimization techniques for simultaneously handling the primal task as well as other performance indices, e.g. quadratic programming. For example, Li et al. identify two limitations of the existing neural network solutions for the quadratic program (QP)-based motion planning of a robot arm, and overcome them by proposing two modified neural network models in [84]. In comparison with an individual robot arm, multi-arm systems are used in many schemes in order to improve performance and reliability. For example, a repetitive motion planning scheme for simultaneously controlling two robot arms is presented in [3], where the two subschemes are unified into a QP-based scheme and then solved by a recurrent neural network. It is an

Kinematic Control of Redundant Robot Arms Using Neural Networks, First Edition.
Shuai Li, Long Jin and Mohammed Aquil Mirza.
© 2019 John Wiley & Sons Ltd. Published 2019 by John Wiley & Sons Ltd.

important issue to tolerate noises online during the end-effector task execution since arms often work in environments with strong electromagnetic interference.

In this chapter, via the NTZNN-based neural-dynamic design method presented in [13], we make progress by presenting a NANTZNN model for the distributed cooperative motion planning of multiple redundant arms.

10.2 Preliminaries, Problem Formulation, and Distributed Scheme

In this section, preliminaries, problem formulation, and distributed scheme are provided to lay a foundation for investigations.

10.2.1 Definition and Robot Arm Kinematics

We present the definition on communication topology of limited communications with $j \in \mathcal{M}(i)$ denoting the neighbor set of the ith redundant robot arm in the communication graph as well as robot arm kinematics in the following [61].

Robot arm kinematics. Given the desired trajectory $r_d(t) \in \mathbb{R}^n$ of the end-effector in work space, we need to generate online the joint trajectory $\theta(t) = [\theta_1(t), \theta_2(t), ..., \theta_m(t)]^T \in \mathbb{R}^m$ in joint space so as to command the robot arm motion. Note that the Cartesian coordinate $r \in \mathbb{R}^n$ in the workspace of a robot arm is uniquely determined by a nonlinear mapping:

$$r(t) = f(\theta(t)), \tag{10.1}$$

where $f(\cdot)$ is a differentiable nonlinear function. Computing time derivations on both sides of (10.1) leads to

$$\dot{r}(t) = J(\theta(t))\dot{\theta}(t), \tag{10.2}$$

where $J(\theta(t)) = \partial f / \partial \theta \in \mathbb{R}^{n \times m}$ is the Jacobian matrix of $f(\theta(t))$, and usually is abbreviated as J. The end-effector $r(t)$ of the redundant robot arm is expected to track the desired path $r_d(t)$, i.e. $r(t) \to r_d(t)$.

10.2.2 Problem Formulation

The constraint for the ith robot arm can be formulated as

$$\sum_{j \in \mathcal{M}(i)} C_{ij}(\varpi_i(t) - \varpi_j(t)) + \delta_i(\varpi_i(t) - r_d(t)) = 0, \tag{10.3}$$

where $\mathcal{M}(i)$ denotes the neighbor set of the ith robot on the communication graph; C_{ij} denotes the connection weight between the ith robot and the jth one; $\varpi_i(t) = r_i(t) - r_{rp}$ with r_{rp} denoting the bias distance between the end-effector and the reference point; $\delta_i = 1$ for $i \in \mathcal{M}(0)$; and $\delta_i = 0$ for $i \notin \mathcal{M}(0)$. Then, the constraint can be formulated as

$$(L \otimes I_n)\overline{\varpi}(t) + (\Psi \otimes I_n)(\overline{\varpi}(t) - \mathbf{1}_p \otimes r_d(t)) = 0, \tag{10.4}$$

where $\otimes$ is the Kronecker product; $\mathbf{1}_p \in \mathbb{R}^{p \times 1}$ denotes a vector composed of 1; $I_n \in \mathbb{R}^{n \times n}$ is an identity matrix; $L = \text{diag}(C\mathbf{1}_p) - C \in \mathbb{R}^{p \times p}$ with $\text{diag}(C\mathbf{1}_p)$ being the diagonal matrix whose p diagonal entries are the p elements of the vector $C\mathbf{1}_p \in \mathbb{R}^{p \times 1}$ with the *ij*th element of C being C_{ij}; $\overline{\omega}(t) = [\omega_1^T(t), \cdots, \omega_p^T(t)]^T \in \mathbb{R}^{np \times 1}$; and $\Psi \in \mathbb{R}^{p \times p}$ is defined as

$$\Psi_{ij} = \begin{cases} \delta_i, & \text{if } i = j, \\ 0, & \text{if } i \neq j. \end{cases}$$

Then, a vector-valued error function could be defined as

$$\xi(t) = (L \otimes I_n)\overline{\omega}(t) + (\Psi \otimes I_n)(\overline{\omega}(t) - \mathbf{1}_p \otimes r_d(t)). \tag{10.5}$$

The following design formula can be used:

$$\dot{\xi}(t) = \gamma \xi(t), \tag{10.6}$$

where design parameter $\gamma > 0$ is used to scale the displacement. Expanding (10.6), we further have

$$(L + \Psi) \otimes \overline{J}(\theta)\dot{\overline{\theta}}(t) = \Psi \otimes I_n \cdot \mathbf{1}_p \otimes \dot{r}_d(t) - \gamma((L \otimes I_n)\overline{\omega}(t)$$
$$+ (\Psi \otimes I_n)(\overline{\omega}(t) - \mathbf{1}_p \otimes r_d(t))), \tag{10.7}$$

where $\overline{\theta}(t) = [\theta_1^T(t), \cdots, \theta_p^T(t)]^T \in \mathbb{R}^{mp \times 1}$ with $\theta_i(t) \in \mathbb{R}^{m \times 1}$ denoting the joint angle in the joint space of the *i*th robot arm and with $\dot{\overline{\theta}}(t)$ denoting its time derivative; and

$$\overline{J}(\theta) = \begin{bmatrix} J_1(\theta_1) & 0 & \cdots & 0 \\ 0 & J_2(\theta_2) & \cdots & 0 \\ \vdots & \vdots & \ddots & \vdots \\ 0 & 0 & \cdots & J_p(\theta_p) \end{bmatrix} \in \mathbb{R}^{np \times mp},$$

with $J_i(\theta_i) \in \mathbb{R}^{n \times m}$ denoting the Jacobian matrix of the *i*th robot arm.

10.2.3 Distributed Scheme

We consider the minimum velocity norm (MVN) performance index in this chapter:

$$U = \sum_{i=1}^{p} \dot{\theta}_i^T(t)\dot{\theta}_i(t)/2 = \| \dot{\overline{\theta}}(t) \|_2^2 /2.$$

Then, the distributed scheme can be formulated as

$$\min \quad \| \dot{\overline{\theta}}(t) \|_2^2 /2, \tag{10.8}$$
$$\text{s.t.} \quad (L + \Psi) \otimes \overline{J}(\theta)\dot{\overline{\theta}}(t) = \Psi \otimes I_n \cdot \mathbf{1}_p \otimes \dot{r}_d(t)$$
$$- \gamma((L \otimes I_n)\overline{\omega}(t) + (\Psi \otimes I_n)(\overline{\omega}(t) - \mathbf{1}_p \otimes r_d(t))).$$

10.3 NANTZNN Solver and Theoretical Analyses

Here, we present a NANTZNN model to solve the MVN-oriented distributed scheme (10.8) with guaranteed convergence and robustness.

10.3.1 NANTZNN for Real-Time Redundancy Resolution

On the basis of [13], we propose and exploit a NANTZNN model for the online solution of the MVN-oriented distributed scheme (10.8).

Solving (10.8) can be done by solving the following equation:

$$\Theta(t)x(t) = \chi(t), \tag{10.9}$$

where

$$\Theta(t) = \begin{bmatrix} I_{mp} & ((L + \Psi) \otimes \bar{J}(\bar{\theta}))^{\mathrm{T}} \\ (L + \Psi) \otimes \bar{J}(\bar{\theta}) & 0 \end{bmatrix},$$

$$x(t) = \begin{bmatrix} \dot{\bar{\theta}}(t) \\ \lambda(t) \end{bmatrix}, \quad \chi(t) = \begin{bmatrix} 0 \\ u(t) \end{bmatrix},$$

with $u = \Psi \otimes I_n \cdot \mathbf{1}_p \otimes \dot{r}_{\mathrm{d}} - \gamma((L \otimes I_n)\overline{\varpi} + (\Psi \otimes I_n)(\overline{\varpi} - \mathbf{1}_p \otimes r_{\mathrm{d}}))$. We define the error function as

$$\zeta(t) = \Theta(t)x(t) - \chi(t). \tag{10.10}$$

To force $\zeta(t)$ to be zero, the following NANTZNN design formula is adopted:

$$\dot{\zeta}(t) = -\varepsilon_1 \Phi(\zeta(t)) - \varepsilon_2 \int_0^t \zeta(\tau)\mathrm{d}\tau, \tag{10.11}$$

where $\varepsilon_1 > 0$ and $\varepsilon_2 > 0$. Besides, $\Phi(\cdot) : \mathbb{R}^n \to \mathbb{R}^n$ denotes a vector array of activation function and the ith element of $\Phi(\cdot)$ is denoted by $\Phi_i(\cdot)$. Generally speaking, any monotonically increasing odd activation function $\Phi(\cdot)$ can be used for constructing the NANTZNN model. In this chapter, the power-sigmoid and hyperbolic sine activation functions are applied in constructing the NANTZNN model:

- The power-sigmoid (ps) activation function (with $\kappa_1 = 3$ and $\varphi = 4$):

$$\Phi_{ips}(e_i) = \begin{cases} \dfrac{1 + \exp(-\varphi)}{1 - \exp(-\varphi)} \dfrac{1 - \exp(-\varphi e_i)}{1 + \exp(-\varphi e_i)}, & \text{if } |e_i| < 1, \\ e_i^{\kappa_1}, & \text{if } |e_i| \geq 1; \end{cases}$$

- The hyperbolic sine (hs) activation function (with $\kappa_2 = 3$):

$$\Phi_{ihs}(e_i) = \frac{\exp(e_i \kappa_2)}{2} - \frac{\exp(-e_i \kappa_2)}{2}.$$

We further obtain the following distributed NANTZNN model:

$$\Theta(t)\dot{x}(t) = -\dot{\Theta}(t)x(t) - \varepsilon_1 \Phi(\Theta(t)x(t) - \chi(t))$$
$$- \varepsilon_2 \int_0^t (\Theta(\tau)x(\tau) - \chi(\tau))\mathrm{d}\tau + \dot{\chi}(t). \tag{10.12}$$

To lay a basis for further investigation on the robustness of MVN-oriented distributed scheme (10.8) aided by distributed NTZNN model (10.12) under the pollution of unknown noises, we have the following equation:

$$\Theta(t)\dot{x}(t) = -\dot{\Theta}(t)x(t) - \varepsilon_1 \Phi(\Theta(t)x(t) - \chi(t))$$
$$- \varepsilon_2 \int_0^t (\Theta(\tau)x(\tau) - \chi(\tau))\mathrm{d}\tau + \dot{\chi}(t) + \sigma(t), \tag{10.13}$$

where $\sigma(t) \in \mathbb{R}^{np+mp}$ denotes the vector-form noises originated from communication noises, computational errors, perturbations, or even their superposition.

10.3.2 Theoretical Analyses and Results

The analyses on the stability, convergence, and robustness of the proposed distributed NANTZNN model (10.12) are conducted in this section using the following theorems.

Theorem 10.1 Distributed NANTZNN model (10.12) is stable and is exponentially and globally convergent to an equilibrium point x^*, of which the first np elements constitute the optimal solution $\bar{\theta}^*$ to the cooperative motion planning of redundant robot arms synthesized by MVN-oriented distributed scheme (10.8).

Proof: It can be generalized from [129] that distributed NANTZNN model (10.12) is stable with $\dot{v}(t) = -2\varepsilon_1\zeta^2(t)$. In addition, we can generalize and obtain the following two facts from Theorem 3 in [5].

- If the power-sigmoid activation function array is exploited, superior convergence is achieved for distributed NANTZNN model (10.12), as compared with the situation of using the linear activation function array presented in [13].
- If the hyperbolic-sine activation function array is exploited, superior convergence to the exponential convergence is achieved for distributed NANTZNN model (10.12), as also compared with the situation of using the power-sigmoid activation function array.

Note that, solving the second-order linear dynamic system $\dot{\zeta}(t) = -\varepsilon_1\zeta(t) - \varepsilon_2\int_0^t \zeta(\tau)\mathrm{d}\tau$ directly, we have the conclusion that, starting from any initial condition $\zeta(0)$, the residual error $\zeta(t)$ of distributed linearly activated NTZNN model for solving MVN-oriented distributed scheme (10.8) globally and exponentially converges to zero. Therefore, we can directly obtain the conclusion that the distributed NANTZNN model (10.12) for solving MVN-oriented distributed scheme (10.8) globally and exponentially converges to zero. That is, the state vector $x(t)$ of distributed NANTZNN model (10.12) globally and exponentially converges to an equilibrium point x^*, of which the first np elements constitute the optimal solution $\bar{\theta}^*$ to the cooperative motion planning of redundant robot arms synthesized by MVN-oriented distributed scheme (10.8). The proof is thus completed. ∎

It is worth investigating the robustness of distributed NANTZNN model (10.12). The following theorem is presented to discuss the performance of distributed NANTZNN model (10.12) with constant noise $\sigma(t) = \varsigma$.

Theorem 10.2 Distributed NANTZNN model (10.12) is stable and is globally convergent to an equilibrium point x^* no matter how large the unknown constant noise $\sigma(t) = \varsigma$ is. Specifically, the vector constituted by the first np elements constitute the optimal solution $\bar{\theta}^*$ to the cooperative motion planning of redundant robot arms synthesized by MVN-oriented distributed scheme (10.8).

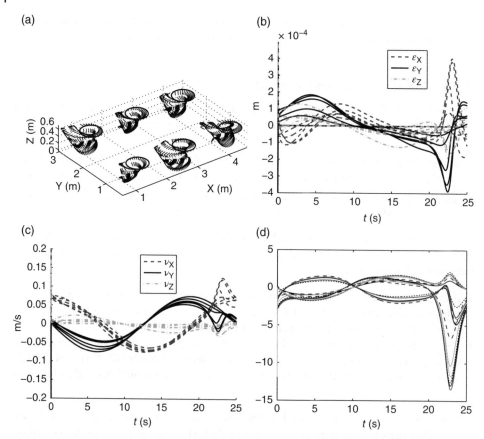

Figure 10.1 Computer simulations synthesized by MVN-oriented distributed scheme (9.8) and distributed NTZNN model (9.13) for consensus of eight redundant PUMA 560 robot manipulators with limited communications and perturbed with constant noise $\rho(t) = 10$, where initial joint states of manipulators are randomly generated. (a) Motion trajectories; (b) profiles of end-effector position errors; (c) profiles of joint angle; and (d) profiles of λ.

Proof: It can be generalized from [13]. ∎

Theorem 10.3 Perturbed with bounded unknown random noise $\sigma(t) = \phi(t) \in \mathbb{R}^{mp+np}$, the cooperative control of redundant robot arms synthesized by MVN-oriented distributed scheme (10.8) and solved by distributed NANTZNN model (10.12) can achieve arbitrary accuracy with sufficiently large γ, sufficiently large ε_1, and with a proper ε_2.

Proof: It can be generalized from [13]. ∎

10.4 Illustrative Examples

In this section, we conduct computer simulations in the situation of zero noise and constant noise.

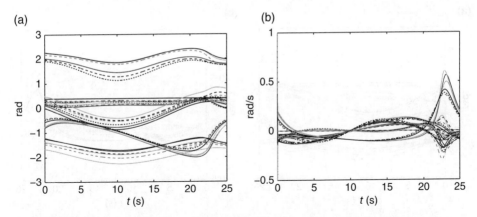

Figure 10.2 Computer simulations synthesized by MVN-oriented distributed scheme (9.8) and distributed NTZNN model (9.13) for consensus of eight redundant PUMA 560 robot manipulators with limited communications and perturbed with constant noise $\rho(t) = 10$, where initial joint states of manipulators are randomly generated. Profiles of (a) joint angle and (b) joint velocity.

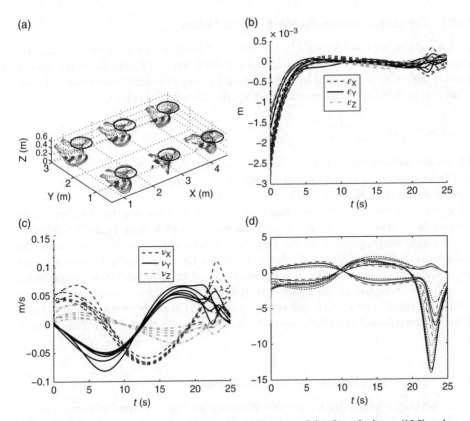

Figure 10.3 Computer simulations synthesized by MVN-oriented distributed scheme (10.8) and hyperbolic-sine activation function activated NANTZNN model (10.12) perturbed with noise $\sigma(t) = 20$ for motion planning of 6 redundant PUMA 560 robot arms with limited communications. (a) Motion trajectories; (b) profiles of end-effector position errors; (c) profiles of joint angle; and (d) profiles of λ.

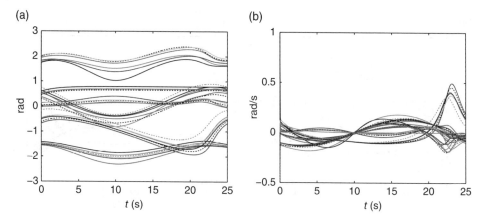

Figure 10.4 Computer simulations synthesized by MVN-oriented distributed scheme (10.8) and hyperbolic-sine activation function activated NANTZNN model (10.12) perturbed with noise $\sigma(t) = 20$ for motion planning of six redundant PUMA 560 robot arms with limited communications. Profiles of (a) joint angle and (b) joint velocity.

10.4.1 Cooperative Motion Planning without Noises

In this section, we consider six PUMA 560 robot arms for cooperative motion planning perturbed without noise. In addition, only robot arm 1 is able to access the desired motion information provided by the command center. The initial joint state of each robot arm is chosen on the desired trajectory. In addition, C_{ij} is set as

$$C_{ij} = \begin{cases} 1, & \text{if } |i - j| \leq 2 \\ 0, & \text{otherwise.} \end{cases}$$

The parameters are chosen as $\gamma = 10, \varepsilon_1 = 100, \varepsilon_2 = 100$, and task duration is $2\Psi/0.25 = 25.12$ s. In addition, the remaining parameters are set as zero (e.g. $\bar{\lambda}$). A typical simulation is generated in Figures 10.1 and 10.2.

It can be observed from Figure 10.3a that all robot arms execute the task denoted by the circular trajectory cooperatively and successfully. In addition, Figure 10.3b shows that the end-effector position tracking errors all remain with a very tiny value during the task execution. The profiles of end-effector velocities, Lagrange-multiplier vector λ, joint angle, and joint velocity are shown in Figures 10.3c, 10.3d, 10.4a, and 10.4b, respectively. It can be observed from these figures that the resulting profiles move smoothly. These simulation results substantiate the effectiveness of MVN-oriented distributed scheme (10.8) and distributed hyperbolic-sine activation function activated NANTZNN model (10.12).

10.4.2 Cooperative Motion Planning with Noises

In this section, we consider cooperative motion planning perturbed with noise $\sigma(t) = 20$. The parameters and communication topology are set the same as those in the previous example with simulation results shown in Figure 10.2. In addition, simulation results based on power-sigmoid activation function activated NANTZNN model (10.12) are presented in Figure 10.5. Note that the detailed descriptions in Figures 10.2 and 10.5 are

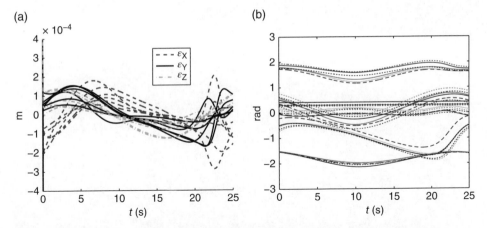

Figure 10.5 Computer simulations synthesized by MVN-oriented distributed scheme (10.8) and power-sigmoid activation function activated NANTZNN model (10.12) perturbed with noise $\sigma(t) = 20$ for motion planning of six redundant PUMA 560 robot arms with limited communications. Profiles of (a) end-effector position errors and (b) joint angle.

omitted due to space limitations. In summary, these simulation results substantiate the effectiveness of MVN-oriented distributed scheme (10.8) and distributed NANTZNN model (10.12).

10.5 Summary

In this chapter, we have proposed a NANTZNN for solving the distributed motion planning of multiple robot arms in the presence of noises. Theoretical analyses have been presented to show the superiorities of the proposed NANTZNN model compared with the existing linearly activated NTZNN model. Furthermore, simulation results have shown the effectiveness and accuracy of the presented distributed scheme with the aid of the NANTZNN model.

References

1 Li, Y., Li, S., and Hannaford, B. (2018). A model based recurrent neural network with randomness for efficient control with applications. *IEEE T. Ind. Inform.* doi: 10.1109/TII.2018.2869588.

2 Zhang, Y., Li, S., Kadry, S., and Liao, B. (2018). Recurrent neural network for kinematic control of redundant manipulators with periodic input disturbance and physical constraints. *IEEE Trans. Cybern.* (99): 1–12. DOI: 10.1109/TCYB.2018.2859751

3 Zhang, Y., Chen, S., Li, S., and Zhang, Z. (2018). Adaptive projection neural network for kinematic control of redundant manipulators with unknown physical parameters. *IEEE Trans. Ind. Electron.* 65 (6): 4909–4920.

4 Jin, L., Zhang, Y., and Li, S. (2016). Integration-enhanced Zhang neural network for real-time-varying matrix inversion in the presence of various kinds of noises. *IEEE Trans. Neural Netw. Learn. Syst.* 27 (12): 2615–2627.

5 Jin, L. and Zhang, Y. (2015). Continuous and discrete Zhang dynamics for real-time varying nonlinear optimization. *Numer. Algorithms* 73 (1): 115–140.

6 Jin, L., Li, S., Hu, B., and Yi, C. (2018). Dynamic neural networks aided distributed cooperative control of manipulators capable of different performance indices. *Neurocomputing* 291: 50–58.

7 Li, S., Wang, H., and Rafique, M. U. (2018). A novel recurrent neural network for manipulator control with improved noise tolerance. *IEEE Trans. Neural Netw. Learning Syst.* 29 (5): 1908–1918.

8 Jin, L., Zhang, Y., Qiao, T. et al. (2016). Tracking control of modified Lorenz nonlinear system using ZG neural dynamics with additive input or mixed inputs. *Neurocomputing* 196: 82–94.

9 Zhang, Y. and Li, S. (2018). A neural controller for image-based visual servoing of manipulators with physical constraints. *IEEE Trans. Neural Netw. Learning Syst.* 29 (11), 5419–5429.

10 Miao, P., Shen, Y., and Xia, X. (2014). Finite time dual networks with a tunable activation function for solving quadratic programming problems and its application. *Neurocomputing* 143: 80–89.

11 Li, S. and Zhang, Y. (2018). *Neural Networks for Cooperative Control of Multiple Robot Arms*. Springer: Singapore.

12 Jin, L. and Li, S. (2017). Nonconvex function activated zeroing neural network models for dynamic quadratic programming subject to equality and inequality constraints. *Neurocomputing* 18: 128–138.

Kinematic Control of Redundant Robot Arms Using Neural Networks, First Edition.
Shuai Li, Long Jin and Mohammed Aquil Mirza.

13 Jin, L. and Li, S. (2018). Distributed task allocation of multiple robots: A control perspective. *IEEE Trans. Syst., Man, Cybern., Syst.* **48** (5): 693–701.

14 Zhang, Z. and Zhang, Y. (2013). Design and experimentation of acceleration-level drift-free scheme aided by two recurrent neural networks. *IET Control Theory Appl.* **7** (1): 25–42.

15 Jin, L., Li, S., La, H.M., and Luo, X. (2017). Manipulability optimization of redundant manipulators using dynamic neural networks. *IEEE Trans. Ind. Electron.* **64** (6): 4710–4720.

16 Li, S., Zhou, M., and Luo, X. (2018). Modified primal-dual neural networks for motion control of redundant manipulators with dynamic rejection of harmonic noises. *IEEE Trans. Neural Netw. Learn. Syst.* **29** (10), 4791–4801.

17 De Silva, C.W. (2007). *Sensors and Actuators: Control System Instrumentation.* Boca Raton, FL: Taylor & Francis.

18 Li, S., Zhou, M., Luo, X., and You, Z.H. (2017). Distributed winner-take-all in dynamic networks. *IEEE Trans. Autom. Control* **62** (2): 577–589.

19 Li, S., Meng, M.Q.H. and Chen, W. (2007). SP-NN: A novel neural network approach for path planning. IEEE International Conference Robotics and Biomimetics, Sanya, Hainan, China.

20 Isidori, A. (1999). *Nonlinear Control Systems II.* New York: Springer-Verlag.

21 Huo,W. (2008). Predictive variable structure control of nonholonomic chained systems. *Int. J. Comput. Math.* **85**: 949–960.

22 Dumitrascu, B., Filipescu, A., Minzu, V., and Filipescu, A. (2007). Backstepping control of wheeled mobile robots. International Conference on System Theory, Control, and Computing, Sinaia, Romania.

23 Bertsekas, D.P. (2005). *Dynamic Programming and Optimal Control.* Nashua, NH: Athena Scientific.

24 Si, J. and Wang, Y.T. (2001). Online learning control by association and reinforcement. *IEEE Trans. Neural Netw.* **12**: 264–276.

25 Hollerbach, J. and Ki, S. (1987). Noise-tolerant ZNN models for solving time-varying zero-finding problems: A Control-Theoretic Approach. *IEEE J. Robotics Automat.* **3** (4): 308–316.

26 O'Neil, K. (2002). Divergence of linear acceleration-based redundancy resolution schemes. *IEEE Trans. Robot. Automat.* **18** (4): 625–631.

27 Kanoun, O., Lamiraux, F., and Wieber, P. (2011). Kinematic control of redundant manipulators: Generalizing the task-priority framework to inequality task. *IEEE Trans. Robot.* **27** (4): 785–792.

28 Escande, A., Mansard, N., and Wieber, P. (2014). Hierarchical quadratic programming: Fast online humanoid-robot motion generation. *Int. J. Robot Res.* **33** (7): 1006–1028.

29 Sadjadian, H., Taghirad, H., and Fatehi, A. (2005). Neural networks approaches for computing the forward kinematics of a redundant parallel manipulator. *Int. J. Comput. Intell.* **2** (1): 40–47.

30 Kumar, N., Panwar, V., Sukavanam, N. et al. (2011). Neural network-based nonlinear tracking control of kinematically redundant robot manipulators. *Math. Comput. Model.* **53** (9–10): 1889–1901.

31 Patchaikani, P., Behera, L., and Prasad, G. (2012). A single network adaptive critic-based redundancy resolution scheme for robot manipulators. *IEEE Trans. Ind. Electron.* **59** (8): 3241–3253.

32 Zhang, S. and Constantinides, A. (1992). Lagrange programming neural networks. *IEEE Trans. Circuits Syst. II* **39** (7): 441–452.

33 Mirza, A. and Li, S. (2015). Dynamic neural networks for kinematic redundancy resolution of parallel Stewart platforms. *IEEE Trans. Cybern.* **99** (8): 3241–3253.

34 Liao, B., Zhang Y., and Jin, L. (2016). Taylor O(h3) Discretization of ZNN models for dynamic equality-constrained quadratic programming with application to manipulators. *IEEE Trans. Neural Netw. Learning Syst.* **27** (2): 225–237.

35 Li, S., Liu, B., Chen, B., and Lou, Y. (2013). Neural network based mobile phone localization using bluetooth connectivity. *Neural Comput. Appl.* **23** (3–4): 667–675.

36 Wang, J. (2010). Analysis and design of a k-winners-take-all model with a single state variable and the heaviside step activation function. *IEEE Trans. Neural Netw.* **21** (9): 1496–1506.

37 Xia, Y., Sun, C., and Zheng, W. (2012). Discrete-time neural network for fast solving large linear L1 estimation problems and its application to image restoration. *IEEE Trans. Neural Netw. Learning Syst.* **23** (5): 812–820.

38 Zhang, Y., Ge, S.S. and Lee, T.H. (2004). A unified quadratic programming based dynamical system approach to joint torque optimization of physically constrained redundant manipulators. *IEEE Trans. Syst. Man Cybern. B Cybern.* **34** (5): 2126–2132.

39 Zhang, Y., Wang, J., and Xia, Y. (2003). A dual neural network for redundancy resolution of kinematically redundant manipulators subject to joint limits and joint velocity limits. *IEEE Trans. Neural Netw.* **14** (3): 658–667.

40 Xia, Y. and Wang, J. (2001). A dual neural network for kinematic control of redundant robot manipulators. *IEEE Trans. Syst., Man, Cybern. B* **31** (1): 147–154.

41 Wang, Z.P., Zhou, T., Mao, Y., and Chen, Q.J. (2014). Adaptive recurrent neural network control of uncertain constrained nonholonomic mobile manipulators. *Int. J. Syst. Sci.* **45** (2): 133–144.

42 Li, S. and Jin, L. (2018). *Competition-Based Neural Networks with Robotic Applications*. Springer: Singapore.

43 Li, S., He, J., Rafique, U., and Li, Y. (2016). Distributed recurrent neural networks for cooperative control of manipulators: a game-theoretic perspective. *IEEE Trans. Neural Netw. Learning Syst.* **14** (3): 658–667.

44 Stengel, R.F. (1994). *Optimal Control and Estimation*, 2e. New York: Dover Publications.

45 Xia, Y., Feng, G., and Wang, J. (2005). A primal-dual neural network for online resolving constrained kinematic redundancy in robot motion control. *IEEE Trans. Syst., Man, Cybern. B* **14** (1): 426–667.

46 Zhang, Y. and Wang, J. (2004). Obstacle avoidance for kinematically redundant manipulators using a dual neural network. *IEEE Trans. Syst., Man, Cybern. B* **34** (1): 752–759.

47 Zhang, Y., Wang, J., and Xu, Y. (2002). A dual neural network for bi-criteria kinematic control of redundant manipulators. *IEEE Trans. Robot. Automat.* **18** (6): 923–931.

48 Khalil, H. (2001). *Nonlinear Systems*, 3e. Upper Saddle River, NJ: Prentice Hall.

49 Liu, G., Xu, J., Wang, X., and Li, Z. (2004). On quality functions for grasp synthesis, fixture planning, and coordinated manipulation. *IEEE Trans. Autom. Sci. Eng.* **1** (2): 146–162.

50 Liu, S. and Wang, J. (2006). A simplified dual neural network for quadratic programming with its KWTA application. *IEEE Trans. Neural Netw.* **17** (6): 1500–1510.

51 Liu, Q. and Wang, J. (2013). A one-layer projection neural network for nonsmooth optimization subject to linear equalities and bound constraints. *IEEE Trans. Neural Netw. Learning Syst.* **24** (5): 812–824.

52 Hu, X. and Wang, J. (2008). An improved dual neural network for solving a class of quadratic programming problems and its-winners-take-all application. *IEEE Trans. Neural Netw.* **19** (12): 2022–2031.

53 Xia, Y., Chen, T., and Shan, J. (2014). A novel iterative method for computing generalized inverse. *Neural Comput.* **26** (2): 449–465.

54 Liu, Q., Guo, Z., and Wang, J. (2012). A one-layer recurrent neural network for constrained pseudoconvex optimization and its application for dynamic portfolio optimization. *Neural Netw.* **26**: 99–109.

55 Xiao, L. and Zhang, Y. (2014). A new performance index for the repetitive motion of mobile manipulators. *IEEE Trans. Cybern.* **44** (2): 280–292.

56 Huang, G. and Cao, J. (2010). Delay-dependent multistability in recurrent neural networks. *Neural Netw.* **23** (2): 201–209.

57 Gao, X. (2003). Exponential stability of globally projected dynamic systems. *IEEE Trans. Neural Netw.* **14** (2): 426–431.

58 Mirza, M.A., Li, S., and Jin, L. (2017). Simultaneous learning and control of parallel Stewart platforms with unknown parameters. *Neurocomputing* **266**: 114–122.

59 Boyd, S. and Vandenberghe, L. (2004). *Convex Optimization*. Cambridge: Cambridge University Press.

60 Zhang, Y., Li, S., and Zhou, X. (2018). Recurrent neural network based velocity-level redundancy resolution for manipulators subject to joint acceleration limit. *IEEE Trans. Ind. Electron.*, DOI: 10.1109/TIE.2018.2851960.

61 Li, S., Kong, R., and Guo, Y. (2014). Cooperative distributed source seeking by multiple robots: algorithms and experiments. *IEEE/ASME Trans. Mech.* **19** (6): 1810–1820.

62 Li, M., Li, W., Niu, L. et al. (2017). An event-related potential-based adaptive model for telepresence control of humanoid robot motion in an environment with cluster obstacles. *IEEE Trans. Ind. Electron.* **64** (2): 1696–1705.

63 Jin, L. and Zhang, Y. (2015). G2-type SRMPC scheme for synchronous manipulation of two redundant robot arms. *IEEE Trans. Cybern.* **45** (2): 153–164.

64 Jin, L., Zhang, Y., Li, S., and Zhang, Y. (2016). Modified ZNN for time-varying quadratic programming with inherent tolerance to noises and its application to kinematic redundancy resolution of robot manipulators. *IEEE Trans. Ind. Electron.* **63** (11): 6978–6988.

65 He, W., Ge, W., Li, Y. et al. (2017). Model identification and control design for a humanoid robot. *IEEE Trans. Syst., Man, Cybern., Syst.* **47** (1): 45–57.

66 La, H.M., Lim, R., and Sheng, W. (2015). Multi-robot cooperative learning for predator avoidance. *IEEE Trans. Contr. Syst. Technol.* **23** (1): 52–63.

67 La, H.M. Lim, R., Basily, B. et al. (2013). Mechatronic and control systems design for an autonomous robotic system for high-efficiency bridge deck inspection and evaluation. *IEEE-ASME Trans. Mechatr.* **18** (6): 1655–1664.

68 La, H.M., Sheng, W., and Chen, J. (2015). Cooperative and active sensing in mobile sensor networks for scalar field mapping. *IEEE Trans. Syst., Man, Cybern. Syst.* **45** (1): 1–12.

69 Nikdel, N., Badamchizadeh, M., Azimirad, V., and Nazari, M.A. (2016). Fractional-order adaptive backstepping control of robotic manipulators in the presence of model uncertainties and external disturbances. *IEEE Trans. Ind. Electron.* **63** (10): 6249–6256.

70 Guo, D. and Zhang, Y. (2014). Li-function activated ZNN with finite-time convergence applied to redundant-manipulator kinematic control via time-varying Jacobian matrix pseudoinversion. *Appl. Soft Comput.* **24**: 158–168.

71 Jin, L. and Zhang, Y. (2014). Discrete-time Zhang neural network of $O(\tau^3)$ pattern for time-varying matrix pseudoinversion with application to manipulator motion generation. *Neurocomputing* **142**: 165–173.

72 Cai, B. and Zhang, Y. (2012). Different-level redundancy-resolution and its equivalent relationship analysis for robot manipulators using gradient-descent and Zhang et al.'s neural-dynamic methods. *IEEE Trans. Ind. Electron.* **59** (8): 3146–3155.

73 Zhang, Y. and Li, S. (2017). Predictive suboptimal consensus of multiagent systems with nonlinear dynamics. *IEEE Trans. Syst., Man, Cybern., Syst.* **47** (7): 1701–1711.

74 Wang, H., Liu, X., and Liu, K. (2016). Robust adaptive neural tracking control for a class of stochastic nonlinear interconnected systems. *IEEE Trans. Neural Netw. Learning Syst.* **27** (3): 510–523.

75 Xiao, L. and Zhang, Y. (2016). Dynamic design, numerical solution and effective verification of acceleration-level obstacle-avoidance scheme for robot manipulators. *Int. J. Syst. Sci.* **47** (4): 932–945.

76 Wang, H., Liu, K., Liu X., and Chen, B. (2015). Neural-based adaptive output-feedback control for a class of nonstrict-feedback stochastic nonlinear systems. *IEEE Trans. Cybern.* **45** (9): 1977–1987.

77 Li, S., Li, Y., Liu, B., and Murray, T. (2012). Model-free control of Lorenz chaos using an approximate optimal control strategy. *Commun. Nonlinear Sci. Numer. Simul.* **17** (12): 4891–4900.

78 Luo, C., Yang, S.X., Li, X., and Meng, M.Q.-H. (2017). Neural dynamics driven complete area coverage navigation through cooperation of multiple mobile robots. *IEEE Trans. Ind. Electron.* **64** (1): 750–760.

79 Na, J., Chen, Q., Ren, X., and Guo, Y. (2014). Adaptive prescribed performance motion control of servo mechanisms with friction compensation. *IEEE Trans. Ind. Electron.* **61** (1): 486–494.

80 Li, S., Li, Y., and Wang, Z. (2013). A class of finite-time dual neural networks for solving quadratic programming problems and its k-winners-take-all application. *Neural Netw.* **39**, 27–39.

81 Li, S., Liu, B., and Li, Y. (2013). Selective positive-negative feedback produces the winner-take-all competition in recurrent neural networks. *IEEE Trans. Neural Netw. Learning Syst.* **24** (2): 301–309.

82 Xia, Y., Feng, G., and Wang, J. (2004). A recurrent neural network with exponential convergence for solving convex quadratic program and related linear piecewise equations. *Neural Netw.* **17** (7): 1003–1015.

83 Zhang, Y., Yan, X., Chen, D. et al. (2016). QP-based refined manipulability maximizing scheme for coordinated motion planning and control of physically constrained wheeled mobile redundant manipulators. *Nonlinear Dyn.* **85** (1): 245–261.

84 Li, S., Zhang, Y., and Jin, L. (2017). Kinematic control of redundant manipulators using neural networks. *IEEE Trans. Neural Netw. Learn. Syst.* **28** (10): 2243–2254.

85 Li, S., Wang, Z., and Li, Y. (2013). Using laplacian eigenmap as heuristic information to solve nonlinear constraints defined on a graph and its application in distributed range-free localization of wireless sensor networks. *Neural Process. Lett.* **37** (3): 411–424.

86 Yoshikawa, T. (1985). Manipulability of robotic mechanisms. *Int. J. Robot Res.* **4** (2): 3–9.

87 Gao, Z. and Zhang, D. (2015). Performance analysis, mapping, and multiobjective optimization of a hybrid robotic machine tool. *IEEE Trans. Ind. Electron.* **62** (1): 423–433.

88 Khalil, H. (1996). *Nonlinear Systems*. Upper Saddle River, NJ: Prentice Hall.

89 Spong, M. and Vidyasagar, M. (2008). *Robot Dynamics and Control*. Hoboken, NJ: Wiley.

90 Liu, S., Li, W., Du, Y., and Fang, L. (2006). Forward kinematics of the Stewart platform using hybrid immune genetic algorithm. *Proceedings of the 2006 IEEE International Conference on Mechatronics and Automation*, Luoyang, China (6–11 June 2006), 2330–2335.

91 Liu, Q., He, Q., and Shi, Z. (2008). Extreme support vector machine classifier. In: *Advances in Knowledge Discovery and Data Mining*, vol. **5012** of Lecture Notes in Computer Science (eds T. Washio, E. Suzuki, K. Ting, and A. Inokuchi), 222–233. Berlin: Springer.

92 Nakamura, Y. and Hanafusa, H. (1986). Inverse kinematics solutions with singularity robustness for robot manipulator control. *ASME J. Dyn. Sys., Meas., Control* **108**: 230–240.

93 Zhang, D. and Lei, J. (2011). Kinematic analysis of a novel 3-DOF actuation redundant parallel manipulator using artificial intelligence approach. *Robot. Comput. Integr. Manuf.* **27** (1): 157–163.

94 Assal, S. (2012). Self-organizing approach for learning the forward kinematic multiple solutions of parallel manipulators. *Robotica* **30** (06): 951–961.

95 Chen, S. and Fu, L. (2013). Output feedback sliding mode control for a Stewart platform with a nonlinear observer-based forward kinematics solution. *IEEE Trans. Control Syst. Technol.* **21**(1): 176–185.

96 Liu, G., Qu, Z., Liu, X., and Han, J. (2014). Tracking performance improvements of an electrohydraulic Gough-Stewart platform using a fuzzy incremental controller. *Ind. Robot* **41** (2): 225C235.

97 Gosselin, C. (1990). Stiffness mapping for parallel manipulators. *IEEE Trans. Robot. Autom.* **6** (3): 377–382.

98 Zhang, Z. and Zhang, Y. (2012). Acceleration-level cyclic-motion generation of constrained redundant robots tracking different paths. *IEEE Trans. Syst., Man, Cybern. B, Cybern.* **42** (4): 1257–1269.

99 Xiao, L. and Zhang, Y. (2013). Acceleration-level repetitive motion planning and its experimental verification on a six-link planar robot manipulator. *IEEE Trans. Control Syst. Technol.* **21** (3): 906–914.

100 Zhang, Z., Li, Z., Zhang, Y. et al. (2015). Neural-dynamic-method-based dualarm CMG scheme with time-varying constraints applied to humanoid robots. *IEEE Trans. Neural Netw. Learning Syst.* **26** (12): 3251–3262.

101 Chen, C.L.P., Wen, G., Liu, Y., and Wang, F. (2014). Adaptive consensus control for a class of nonlinear multiagent time-delay systems using neural networks. *IEEE Trans. Neural Netw. Learning Syst.* **25** (6): 1217–1226.

102 Maeso, S., Reza, M., Mayol, J. et al. (2010). Efficacy of the Da Vinci surgical system in abdominal surgery compared with that of laparoscopy: A systematic review and meta-analysis. *Ann. Surg.* **252** (2): 254–262.

103 Aiyama, Y., Hara, M., Yabuki, T. et al. (1999). Cooperative transportation by two four-legged robots with implicit communication. *Robot. Auton. Syst.* **29** (1): 13–19.

104 Welch, J., Backer, D., Bauermeister, A. et al. (2009). The Allen telescope array: The first widefield, panchromatic, snapshot radio camera for radio astronomy and SETI. *Proc. IEEE* **97** (8): 1438–1447.

105 Jin, L. and Li, S. La, H., Zhang, X., and Hu, B. (2019). Distributed task allocation in multi-robot coordination for moving target tracking: a distributed approach. *Automatica*.

106 Chen, C.L.P., Wen, G., Liu, Y., and Liu, Z. (2016). Observer-based adaptive back-stepping consensus tracking control for high-order nonlinear semi-strict-feedback multiagent systems. *IEEE Trans. Cybern.* **46** (7): 1591–1601.

107 Wen, G., Chen, C.L.P., Liu, Y., and Liu, Z. (2015). Neural-network-based adaptive leader following consensus control for second-order non-linear multi-agent systems. *IET Control Theory Appl.* **9** (13): 1927–1934.

108 Wang, D., Zhang, N., Wang, J., and Wang, W. (2017). Cooperative containment control of multiagent systems based on follower observers with time delay. *IEEE Trans. Syst., Man, Cybern., Syst.* **47** (1): 13–23.

109 Liu, Y. and Tong, S. (2015). Adaptive fuzzy identification and control for a class of nonlinear pure-feedback MIMO systems with unknown dead zones. *IEEE Trans. Fuzzy Syst.* **23** (5): 1387–1398.

110 Liu, Z.-W., Yu, X., Guan, Z.-H. et al. (2017). Pulse-modulated intermittent control in consensus of multiagent systems. *IEEE Trans. Syst., Man, Cybern., Syst.* **47** (5): 783–793.

111 Zhang, Y. and Li, S., and Jiang, X. (2018). Near-optimal control without solving HJB equations and its applications. *IEEE Trans. Ind. Electron.* **65** (9): 7173–7184.

112 He, W., Chen, Y., and Yin, Z. (2016). Adaptive neural network control of an uncertain robot with full-state constraints. *IEEE Trans. Cybern.* **46** (3): 620–629.

113 He, W., Dong, Y., and Sun, C. (2016). Adaptive neural impedance control of a robotic manipulator with input saturation. *IEEE Trans. Syst., Man, Cybern., Syst.* **46** (3): 334–344.

114 Wang, H., Chen, B., Liu, K. et al. (2015). Adaptive neural tracking control for a class of nonstrict-feedback stochastic nonlinear systems with unknown backlash-like hysteresis. *IEEE Trans. Neural Netw. Learning Syst.* **25** (5): 947–958.

115 Li, S., Guo, Y., and Bingham, B. (2014). Multi-robot cooperative control for monitoring and tracking dynamic plumes. *2014 IEEE International Conference on Robotics and Automation (ICRA)*, 67–73. IEEE.

116 Wang, H., Shi, P., Li, H., and Zhou, Q. (2017). Adaptive neural tracking control for a class of nonlinear systems with dynamic uncertainties. *IEEE Trans. Cybern.* **47** (10): 3075–3087.

117 Zhang, Y. and Yi, C. (2011). *Zhang Neural Networks and Neural-Dynamic Method*. Hauppauge, NY: Nova.

118 Jin, L., Zhang, Y., Li, S., and Zhang, Y. (2017). Noise-tolerant ZNN models for solving time-varying zero-finding problems: A Control-Theoretic Approach. *IEEE Trans. Autom. Control* **62** (2): 992–997.

119 Jin, L. and Zhang, Y. (2015). Discrete-time Zhang neural network for online time-varying nonlinear optimization with application to manipulator motion generation. *IEEE Trans. Neural Netw. Learning Syst.* **26** (7): 1525–1531.

120 Li, S., Chen, S., and Liu, B. (2013). Accelerating a recurrent neural network to finite-time convergence for solving time-varying Sylvester equation by using a sign-bi-power activation function. *Neural Process. Lett.* **37** (2): 189–205.

121 Xiao, L. and Liao, B. (2016). A convergence-accelerated Zhang neural network and its solution application to Lyapunov equation. *Neurocomputing* **193**: 213–218.

122 Guo, D., Nie, Z., and Yan, L. (2016). Theoretical analysis, numerical verification and geometrical representation of new three-step DTZD algorithm for time-varying nonlinear equations solving. *Neurocomputing* **214**: 516–526.

123 Guo, D., Nie, Z., and Yan, L. (2017). Novel discrete-time Zhang neural network for time-varying matrix inversion. *IEEE Trans. Syst., Man, Cybern., Syst.* **47** (8): 2301–2310.

124 Zhang, Y., Lv, X., Li, Z. et al. (2008). Effective neural remedy for drift phenomenon of planar three-link robot arm using quadratic performance index. *Electron. Lett.* **44** (6): 1–2.

125 Li, S., Wang, Y., Yu, J., and Liu, B. (2013). A nonlinear model to generate the winner-take-all competition. *Commun. Nonlinear Sci. Numer. Simul.* **18** (3): 435–442.

126 Li, S., Chen, S., Liu, B. et al. (2012). Decentralized kinematic control of a class of collaborative redundant manipulators via recurrent neural networks. *Neurocomputing* **91**: 1–10.

127 Li, S., Cui, H., Li, Y. et al. (2012). Decentralized control of collaborative redundant manipulators with partial command coverage via locally connected recurrent neural networks. *Neural Comput. Appl.* **23** (3): 1051–1060.

128 Jin, L., Li, S., Liao. B., and Zhang, Z. (2017). Zeroing neural networks: A survey. *Neurocomputing* **267** (6): 597–604.

129 Jin, L., Li, S., Xiao, L. et al. (2018). Cooperative motion generation in a distributed network of redundant robot manipulators with noises. *IEEE Trans. Syst., Man, Cybern., Syst.* **48** (10): 1715–1724.

Index

Kinematic Control of Redundant Robot Arms Using Neural Networks, First Edition.
Shuai Li, Long Jin and Mohammed Aquil Mirza.
© 2019 John Wiley & Sons Ltd. Published 2019 by John Wiley & Sons Ltd.